AF784675

CONTENTS

MAY 23, 2006

OPENING STATEMENT

WITNESSES

APPENDIX

Material Submitted for the Hearing Record

ANIMAL ENTERPRISE TERRORISM ACT

HEARING

BEFORE THE

SUBCOMMITTEE ON CRIME, TERRORISM, AND HOMELAND SECURITY

OF THE

COMMITTEE ON THE JUDICIARY HOUSE OF REPRESENTATIVES

ONE HUNDRED NINTH CONGRESS

SECOND SESSION

ON

H.R. 4239

MAY 23, 2006

Serial No. 109–125

Printed for the use of the Committee on the Judiciary

Available via the World Wide Web: http://judiciary.house.gov

U.S. GOVERNMENT PRINTING OFFICE

27–742 PDF WASHINGTON : 2006

For sale by the Superintendent of Documents, U.S. Government Printing Office
Internet: bookstore.gpo.gov Phone: toll free (866) 512–1800; DC area (202) 512–1800
Fax: (202) 512–2250 Mail: Stop SSOP, Washington, DC 20402–0001

COMMITTEE ON THE JUDICIARY

F. JAMES SENSENBRENNER, JR., Wisconsin, *Chairman*

HENRY J. HYDE, Illinois
HOWARD COBLE, North Carolina
LAMAR SMITH, Texas
ELTON GALLEGLY, California
BOB GOODLATTE, Virginia
STEVE CHABOT, Ohio
DANIEL E. LUNGREN, California
WILLIAM L. JENKINS, Tennessee
CHRIS CANNON, Utah
SPENCER BACHUS, Alabama
BOB INGLIS, South Carolina
JOHN N. HOSTETTLER, Indiana
MARK GREEN, Wisconsin
RIC KELLER, Florida
DARRELL ISSA, California
JEFF FLAKE, Arizona
MIKE PENCE, Indiana
J. RANDY FORBES, Virginia
STEVE KING, Iowa
TOM FEENEY, Florida
TRENT FRANKS, Arizona
LOUIE GOHMERT, Texas

JOHN CONYERS, JR., Michigan
HOWARD L. BERMAN, California
RICK BOUCHER, Virginia
JERROLD NADLER, New York
ROBERT C. SCOTT, Virginia
MELVIN L. WATT, North Carolina
ZOE LOFGREN, California
SHEILA JACKSON LEE, Texas
MAXINE WATERS, California
MARTIN T. MEEHAN, Massachusetts
WILLIAM D. DELAHUNT, Massachusetts
ROBERT WEXLER, Florida
ANTHONY D. WEINER, New York
ADAM B. SCHIFF, California
LINDA T. SÁNCHEZ, California
CHRIS VAN HOLLEN, Maryland
DEBBIE WASSERMAN SCHULTZ, Florida

PHILIP G. KIKO, *General Counsel-Chief of Staff*
PERRY H. APELBAUM, *Minority Chief Counsel*

SUBCOMMITTEE ON CRIME, TERRORISM, AND HOMELAND SECURITY

HOWARD COBLE, North Carolina, *Chairman*

DANIEL E. LUNGREN, California
MARK GREEN, Wisconsin
TOM FEENEY, Florida
STEVE CHABOT, Ohio
RIC KELLER, Florida
JEFF FLAKE, Arizona
MIKE PENCE, Indiana
J. RANDY FORBES, Virginia
LOUIE GOHMERT, Texas

ROBERT C. SCOTT, Virginia
SHEILA JACKSON LEE, Texas
MAXINE WATERS, California
MARTIN T. MEEHAN, Massachusetts
WILLIAM D. DELAHUNT, Massachusetts
ANTHONY D. WEINER, New York

MICHAEL VOLKOV, *Chief Counsel*
DAVID BRINK, *Counsel*
CAROLINE LYNCH, *Counsel*
JASON CERVENAK, *Full Committee Counsel*
BOBBY VASSAR, *Minority Counsel*

ANIMAL ENTERPRISE TERRORISM ACT

TUESDAY, MAY 23, 2006

HOUSE OF REPRESENTATIVES,
SUBCOMMITTEE ON CRIME, TERRORISM,
AND HOMELAND SECURITY
COMMITTEE ON THE JUDICIARY,
Washington, DC.

The Subcommittee met, pursuant to notice at 10 a.m., in Room 2141, Rayburn House Office Building, the Honorable Howard Coble (Chairman of the Subcommittee) presiding.

Mr. COBLE. Good morning, ladies and gentlemen.

Today the Subcommittee on Crime, Terrorism, and Homeland Security is conducting a legislative hearing on H.R. 4239, the "Animal Enterprise Terrorism Act," which was introduced on November 4, 2005, by several of our colleagues. And the lead sponsor is the distinguished gentleman from Wisconsin Mr. Petri, whom I recognize in the audience today, along with Chairman Sensenbrenner, and Representatives Issa, McCotter, Cannon, Bonilla, Calvert, Otter, Boren, Blackburn and Doolittle.

H.R. 4239 was introduced in response to a growing threat commonly referred to as ecoterrorism. While we are still responding to the threat about international terrorism, groups of impassioned animal supporters have unfortunately employed tactics to disrupt animal research and related businesses by terrorizing their employees. Today's testimony will detail what employees have come to fear, but it is safe to say that their fear is real and justified.

This practice originated with protests against companies conducting animal research. The protests became violent, and as they continue in severity, they are now being focused on employees of businesses with any remote relationship to the primary research. The range of potential victims includes employees of banking, insurance, securities and pharmaceutical companies, and even universities.

Dr. Tom O'Connor of North Carolina Wesleyan College teaches a course on the different types of terrorism. According to Dr. O'Connor, and I quote, "Ecoterrorism involves extremist views on environmental issues and animal rights, and is a fringe-issue form of terrorism aimed primarily at inflicting economic damage on those seen as profiting from the destruction and exploitation of the environment," closed quote. Dr. O'Connor distinguishes the environmentalist movement from the more extreme ecoterrorists in this way, and again I quote, quote, "Environmentalists work within the system for preservation, and ecoterrorists seem to want to destroy civilization as we know it in order to save the planet," closed quote.

Because many of these acts are not considered an offense under the current animal enterprise terrorism statute, that is, 18 U.S. Code 43, H.R. 4239 would expand the reach of the animal enterprise terrorism statute to specifically include the use of force, violence or threats against entities that do business with animal enterprise organizations. Specifically, the legislation would prohibit the international damaging of property—or strike that—the intentional damaging of property of a person or entity having a connection to, relationship with or transactions with an animal enterprise, and make it a criminal act to intentionally place a person in reasonable fear of death or serious bodily injury to that person or his or her family because of his or her relationship with the animal enterprise.

Since the bill has been introduced, the Committee has been approached by a couple of groups with concerns about ensuring first amendment protections that are included for lawful protests, boycotts and other activities. The legislation was not intended to infringe on these rights in any way. Accordingly, a manager's amendment clarifying that those rights will continue to be protected was included in Members' packets and will be introduced at a subsequent markup on which Members can cast their votes.

I have received numerous statements to be entered into the record in support of this bill, including statements from the House and Senate sponsors of this legislation, Representative Petri, who I mentioned earlier, and Senator Inhofe, the distinguished gentleman from Oklahoma.

And I look forward to hearing from the witnesses today, and especially would like to thank Dr. Michele Basso and Bill Trundley for their willingness to testify about their experiences.

[The prepared statement of Mr. Petri follows in the Appendix]

[The prepared statement of Senator Inhofe follows in the Appendix]

Mr. COBLE. At this time I am pleased to recognize the distinguished gentleman from Virginia, the Ranking Member of this Subcommittee, Mr. Bobby Scott.

Mr. SCOTT. Thank you, Mr. Chairman. I would like to thank you for holding a markup on H.R. 4239, the "Animal Enterprise Terrorism Act." Apparently, our current Federal law that was designed to protect businesses and employees in animal enterprises are doing a pretty good job. However, many of these businesses and employees are now complaining that other businesses and nonprofits and their employees, board members and family members with whom they are affiliated are being stalked, harassed, intimidated. They have had their businesses, homes or cars vandalized, and some individuals even physically assaulted.

Indications are that animal rights groups that have used extreme tactics to press their point of view were taking advantage of the fact that animal enterprise laws do not cover these types of secondary relationships to wage a campaign of threats, harassment, intimidation and fear-mongering in an effort to have them sever their relationships with targeted animal enterprises. This bill was designed to cover these perceived gaps or loopholes in the current animal enterprise protection laws.

Citizens engaging in lawful activities as well as those associated with them are entitled to be protected from criminal acts and to be able to go about their lawful activities free from threats to their person or property and that of their family and associates. State laws are generally good at providing those protections. However, the interstate nature of the planning and conduct of these criminal and harassment tactics by groups skilled at avoiding the laws make it difficult for States to effectively get at some of the problems, and that is what the bill is designed to cover.

While we must protect those engaged in lawful animal enterprises, we must also protect the right of those engaged in their first amendment freedoms and expressions regarding such enterprises. The issue was acknowledged and addressed in the bill. However, we received concerns that protections do not go far enough to ensure that first amendment freedoms are not compromised.

I am aware, Mr. Chairman, of your proposals to further improvements in this area, and I want to work with you to ensure that we provide the protections of lawful activities that are needed here without jeopardizing first amendment freedoms. Included in those freedoms, Mr. Chairman, is a right to engage in peaceful civil disobedience, and I'm not sure the proposals adequately take that into account. If a group's intention were to stage a sit-down, lie-down or to block traffic to a targeted facility, they certainly run the risk of arrest for whatever traffic, trespass or other laws they are breaking, but they should not be held any more accountable for business losses due to delivery trucks being delayed any more than anyone else guilty of such activities.

I am concerned, Mr. Chairman, that the bill treats conspiracies and attempts the same as a completion of an offense. While someone who has not completed an offense solely because they were caught by law enforcement before the completion should not be rewarded, I believe we should also encourage potential offenders to change their mind at any time. Insisting that offenders who decide not to go through with an offense will get the same sentence as if they had only helps ringleaders or others promote the philosophy that if I am going to be shot for being a wolf, I might as well eat the sheep.

So, Mr. Chairman, I look forward to testimony by witnesses to see how we can strike a proper balance between protecting lawful activities and our first amendment freedoms. Thank you.

Mr. COBLE. I thank the gentleman.

[The information referred to follows in the Appendix]

Mr. COBLE. Lady and gentlemen, it is the practice of the Subcommittee to swear in all witnesses appearing before it, so if you would, please, stand and raise your right hands.

[Witnesses sworn.]

Mr. COBLE. Let the record show that the witnesses answered in the affirmative. And you may be seated.

We have a distinguished panel today. Ladies and gentlemen, we are glad to welcome the rest of you in the audience as well. Our first witness is Deputy Assistant Attorney General Brent McIntosh. Mr. McIntosh is a Deputy Assistant Attorney General in the United States Department of Justice Office of Legal Policy. Prior to joining the Justice Department, Mr. McIntosh was an attorney with the New York law firm of Sullivan and Cornwell. He also served as a law clerk for the United States Court of Appeals for the D.C. Circuit and the U.S. Circuit Court of Appeals for the Second Circuit.

Mr. McIntosh was awarded an undergraduate degree from the University of Michigan and holds a J.D. from Yale University.

Our second witness is Dr. Michele Basso. Dr. Basso is an assistant professor with the University of Wisconsin's Department of Physiology, Ophthalmology and Visual Sciences. Dr. Basso's current emphasis is on understanding normal brain mechanisms controlling complex behaviors and how these mechanisms go awry in movement-disordered states. This research seeks to reveal the neurophysiological underpinnings of movement disorders such as Parkinson's disease, Huntingdon's disease and dystonia. Her research is highly coordinated with practicing clinicians, and in some instances uses animal models, in particular the rhesus monkey. Dr. Basso received her doctorate from SUNY in Stony Brook.

Ophthalmologist, I just wanted to make sure I can say it, Doctor.

Our third witness is Mr. William Trundley, Vice President of Corporate Security and Investigations for GlaxoSmithKline. Mr. Trundley has global responsibility for a range of areas, including countering extremist activity against the company and its employees, product security, protection of personnel information and assets, security risk analysis and investigations.

Mr. Trundley has served for 24 years as a major with the Royal Military Police Special Investigation Branch during which time he undertook a variety of assignments in several overseas areas, including Europe, North America and the Far East. Mr. Trundley holds an M.S. from Western University and a United States diploma in security management.

Our final witness today, Mr. William Potter. Mr. Potter is a freelance reporter based in Washington, D.C., and has focused attention on animal rights and environmental activists whose activities result in prosecutions and the civil rights implications involved. He has written for publications including the *Chicago Tribune, The Dallas Morning News, Legal Affairs, The Chronicle of Higher Education, In These Times, The Texas Observer, The Washington City Paper, Z and CounterPunch.* Mr. Potter was graduated summa cum laude from the University of Texas at Austin with a degree in journalism.

We've been privileged to be joined by the distinguished gentleman from Massachusetts Mr. Delahunt.

So Mr. McIntosh, why don't you be our lead-off hitter today, and we will let Mr. Potter be the clean-up hitter. And as we have told you all previously, folks, Mr. Scott and I try to operate under the 5-minute rule. When the amber light appears before you on the panel, that is your warning that you have 1 minute remaining. Now you will not be keel-hauled if you fail to conclude in 5 minutes. When the red light appears, that's your warning that 5 minutes have elapsed, and if you could wrap up at that point.

Mr. McIntosh.

TESTIMONY OF BRENT McINTOSH, DEPUTY ASSISTANT ATTORNEY GENERAL, UNITED STATES DEPARTMENT OF JUSTICE

Mr. McINTOSH. Thank you, Mr. Chairman, Ranking Member Scott, Congressmen. Good morning.

On behalf of the Department of Justice, I thank you for this opportunity to testify regarding the Department's efforts to investigate and prosecute those who threaten violence and commit criminal acts in the name of protecting animals.

The Department remains dedicated to protecting the American people from the threat of violence imposed by extremists, while at the same time protecting the first amendment rights guaranteed to all Americans.

We have had some success in prosecuting animal rights extremists. Most recently, on March 2, 2006, six members of an animal rights group called Stop Huntingdon Animal Cruelty, or SHAC, were convicted of inciting attacks on those who worked for or did business with Huntingdon Life Sciences, a British company that runs an animal testing laboratory.

As demonstrated by the SHAC convictions, extremists have not hesitated to use violence and the threat of violence to further their social and political goals. In some cases, such as those involving arson or explosives, Federal prosecutors are well equipped to prosecute and punish extremists, but not all animal rights extremists use arson and explosives.

In pursuit of its goal of closing Huntingdon Life Sciences' animal testing operations, SHAC and its sympathizers have employed a wide variety of harassing and intimidating techniques which SHAC itself calls its, quote, "top 20 terror tactics," end quote, designed to terrorize SHAC's targets while avoiding an effective law enforcement response. For example, these violent extremists have advocated and facilitated such direct actions as vandalizing—including fire-bombing homes, businesses and cars—fraud and ID theft; making bomb threats or threats to harm or kill targets, targets' partners, targets' children.

To target these techniques, SHAC has posted on the Internet law-abiding employees' home telephone numbers, the names of their spouses and children, even the schools where those children attend. In short, these extremists are engaged in a nationwide campaign to place law-abiding citizens in a reasonable fear of death or of serious bodily injury to themselves or their loved ones.

Although the existing Animal Enterprise Protection Act is an important tool for prosecutors, animal rights extremists have tailored their campaigns to exploit limits and ambiguities in the statute by targeting individuals and businesses associated with the animal enterprise rather than the animal enterprise itself. Considered individually, these actions are State crimes, but local police often lack the investigative resources and nationwide perspective to put these local offenses into context as a multijurisdictional campaign of violence. So while the U.S. Attorney in New Jersey can prosecute some of SHAC's crimes under the existing statute, most of the charges brought in that prosecution came under the interstate stalking statute.

The bill under consideration today would fill gaps in the current law, and the Department supports it. Most important, as the Ranking Member said, the existing statute's focus on physical disruption of the actual animal enterprise permits the argument that it does not cover a campaign that harms the animal enterprise, not directly, but by targeting persons and entities that do business with an animal enterprise. H.R. 4239 would make clear that committing the proscribed conduct against an employee of an animal enterprise or against an entity related to an animal enterprise is equally illegal.

Before I conclude, let me spend a moment on people the Department does not prosecute. The Department is acutely aware of the importance of protecting the first amendment rights of those who lawfully protest the treatment of animals. Let me say this as clearly as I can: The Department does not prosecute and does not wish to prosecute those who lawfully seek to persuade others. On this issue the Department has found wide common ground with members of the Humane Society and the ACLU. We recently met with both groups. We all agree that any tactic or strategy of involving violence or threats of violence is not to be tolerated. On the other hand, we are committed to ensuring that the law has no chilling effect on lawful activities designed merely to persuade.

This proposed law builds on existing concepts in the Federal Criminal Code, and as a legal matter breaks no new ground. Still, the Department has heard the concerns of the Humane Society and the ACLU, has seen the manager's amendment, and is happy to work with the Subcommittee to leave no doubt that nothing in the law prohibits any expressive conduct protected by the first amendment.

The great majority of animal rights advocates make their case through lawful first amendment activity, but those who cross the line from free speech to criminal conduct should be prosecuted and punished appropriately, and prosecutors should have the tools to make sure that happens.

Mr. Chairman, thank you again for inviting us here today. We thank this Subcommittee for its continued leadership and support, and we welcome your questions.

Mr. COBLE. Thank you, Mr. McIntosh. And your timing was superb, you ended at the right time with the red light.

[The prepared statement of Mr. McIntosh follows:]

PREPARED STATEMENT OF BRENT J. MCINTOSH

Testimony of
Brent J. McIntosh
Deputy Assistant Attorney General
Office of Legal Policy
U.S. Department of Justice

Legislative Hearing on H.R. 4239
"The Animal Enterprise Terrorism Act"

before the
Subcommittee on Crime, Terrorism and Homeland Security
U.S. House of Representatives

May 23, 2006

Chairman Coble and members of the Subcommittee, thank you for giving me the opportunity to appear here today and testify before you on this important subject. I am pleased to discuss the Department's efforts to investigate and prosecute entities and individuals who threaten violence and commit criminal acts against our fellow citizens in the name of protecting animals. These matters are an important part of the mission of the Department of Justice to protect the American people from acts of violence and threats of violence.

As you know, the safety and security of the American people is the number one priority of the Department of Justice. As such, we remain dedicated to the task of protecting them from violence and the threat of violence posed by extremists while at the same time protecting the First Amendment and other civil liberties guaranteed to all Americans in the Constitution.

Successes in the Fight Against Domestic Extremists

The Department has had some successes in prosecuting animal rights extremists who have violated federal law. In the Western District of Wisconsin, Peter Young pleaded guilty on September 2, 2005, to violations of the Animal Enterprise Protection Act arising from the 1998 release of minks from farms in Wisconsin. Most recently, on March 2, 2006, six members of an animal rights group called Stop Huntingdon Animal Cruelty (or SHAC) were convicted of terrorism and Internet stalking by a federal jury that found them guilty of using a website to incite attacks on those who did business with or worked for a British company that runs an animal testing laboratory in New Jersey.

The Threat Posed By Animal Rights Extremists

As this Subcommittee well knows, extremists have not hesitated to use violence and the threat of violence to further their social and political goals. In cases in which individuals have used improvised incendiary or explosive devices, federal prosecutors are well-equipped to prosecute and punish such individuals using the tools provided in Title 18, United States Code section 844. That statute provides for substantial penalties for those who use fire or explosives to damage property that falls within the scope of the statute. For violations of subsections (f) and (i) – the two sections most frequently at issue in such cases – defendants face a mandatory minimum sentence of 5 years in prison and a maximum sentence of 20 years. Section 924(c) of Title 18 – which proscribes, among other things, using a destructive device during the commission of a federal crime of violence – can lead to additional mandatory prison time up to life.

Violence by animal rights extremists is not limited, however, to the use of arson and explosives. Stop Huntingdon Animal Cruelty and other like-minded animal rights extremist entities are engaging in a campaign of criminal conduct that can only be described as terrorizing those whom they identify as targets. In pursuit of its goal of closing Huntingdon Life Science's animal testing operations, SHAC's campaign included a wide variety of harassing and terrifying techniques specifically designed to terrorize the subjects of those efforts while avoiding an effective law enforcement response. In fact, the SHAC website promoted a "Top 20 Terror Tactics" targeting places and persons. The personal and economic consequences of this campaign have been, and will continue to be, significant.

Tools for the Prosecution of Animal Rights Extremists

The Animal Enterprise Protection Act, codified at section 43 of Title 18, is an important tool for prosecutors seeking to combat animal rights extremists. This statute was passed in 1992 primarily to address the problem of those who physically intruded upon the property of entities who tested or otherwise used animals and then damaged the property belonging to the animal enterprise. Originally established as a misdemeanor, the statute's penalties have been enhanced by amendments in 1996 and 2002.

While this statute is an important tool for prosecutors, SHAC and other animal rights extremists have recognized limits and ambiguities in the statute and have tailored their campaign to exploit them by violent acts targeting individuals and business that do not "physically disrupt" the animal enterprise itself. For example, these violent extremists have advocated such "direct actions" as:

- Spray painting abusive graffiti and vandalizing one's home, business or car;
- Physically assaulting associated individuals, including spraying cleaning fluid into eyes;
- Smashing home windows while one's family is present;

- Making false bomb threats and threatening telephone calls and letters, including threats to kill or injure one's partner or children;

- Internet posting of the home telephone numbers of law-abiding employees with the names of their spouses and children along with their names, ages, birth dates and schools where these children attend or license plate numbers of the employees' cars and where they attend church;

In short, these extremists are engaged in a nationwide campaign to intentionally place our fellow law-abiding citizens in reasonable fear of the death of, or serious bodily injury to, themselves or loved ones because of their association with animal enterprises. However, there are currently no federal criminal laws that directly facilitate the investigation and prosecution of such outrageous, violent acts. These extremists are traveling in interstate commerce or using the mail or facilities of such commerce to commit what are now state crimes involving threats, acts of vandalism, property damage, trespass, harassment or intimidation. However, county prosecutors and local police often do not have the investigative resources, expertise, or nationwide perspective to put these types of local offenses in the context of a multi-jurisdictional campaign of violence for the purpose of damaging or disrupting an animal enterprise.

With no federal hook, the FBI and federal prosecutors are finding it difficult to open grand jury investigations and bring the resources and expertise of the United States to bear on this problem. While the U.S. Attorney in New Jersey could skillfully prosecute some of the SHAC crimes under the existing statue, he nevertheless was limited by charging other conduct under the interstate stalking statute in 18 U.S.C. § 2261A which was designed to protect women from domestic violence. This engendered extensive motion practice in which the attorneys for the government had to defend the indictment from numerous legal attacks. Moreover, under some circumstances, other districts are finding it difficult because many cases do not fall within section 43 under the theories used in the SHAC prosecution.

The bill under consideration by the Subcommittee would fill the gaps in the current law and enable federal law enforcement to investigate and prosecute these violent felonies. The proposed amendment of the statute will make clear and unequivocal the application of the statute to recent trends in animal rights extremism and will enhance the effectiveness of the Department's response to this domestic threat.

Proposed Amendment of Section 43, Title 18, United States Code

For this reason, the Department supports H.R. 4239 to amend the Animal Enterprise Protection Act in order to address several infirmities that keep prosecutors from using it in the most effective manner possible.

First, the statute's definition of the type of "animal enterprise" that it protects is not broad enough to include some of the entities that are now targeted by animal rights extremists. These include pet stores and even animal shelters. The threat posed to individuals associated with such organizations is no less significant than the threat that gave rise to the original statute. The bill under consideration by this Subcommittee would expand the definition of "animal enterprise" so that such entities are clearly included within the scope of the statute.

Second, the existing statute's use of the phrase "physical disruption" to describe the conduct it proscribes unnecessarily suggests that it covers a narrow scope of conduct tantamount to trespass. In that regard, the existing statute permits the argument that it does not cover actions by SHAC or other animal rights extremists taken not against an animal enterprise, but against those persons and other entities that choose to do business with an animal enterprise. This proposal avoids the narrowness of "physical disruption" by focusing instead on economic damage and disruption resulting from threats, acts of vandalism, property damage, trespass, harassment or intimidation and the reasonable fear of death or serious bodily injury occasioned by such criminal actions. Such a law would more effectively protect animal enterprises and their employees from the criminal conduct in which animal rights extremists such as SHAC currently engage.

Finally, in its current form, the statute fails to address clearly the consequences of a campaign of vandalism and harassment directed against individuals – as opposed to the animal enterprise itself. H.R. 4239 would remedy this ambiguity by clearly stating that committing the proscribed conduct against an employee of an animal enterprise (or of an entity with a relationship with an animal enterprise) is equally illegal.

Taken together, the changes in H.R. 4239 would empower prosecutors with a more effective tool to meet the challenges now posed by these animal rights extremists. We strongly encourage the Subcommittee to endorse this proposal.

<u>First Amendment</u>

In seeking to meet the challenge of these changing forms of criminal conduct by animal rights extremists, the Department is acutely aware of the importance of protecting the First Amendment rights of those who protest the testing and other use of animals. Let me say this as clearly as I can: The Department does not seek to prosecute those who enter the arena of debate seeking to persuade their government or private businesses and individuals of the merit of their viewpoints.

On this issue the Department has found wide common ground with members of the Humane Society and the ACLU. We all agree that any tactic or strategy involving violence toward people or threats of violence is wholly unacceptable and inconsistent with a core ethic of promoting compassion and respect. Such violence is wrong and not to be defended or tolerated. We all agree that more must be done to crack down on this violence, no matter what cause it hides behind. Moreover, we all are working towards a bill that would have no chilling effect on legal, mainstream activities that should be part of the public discourse in this country. We do not want to prohibit or discourage the protected activities of whistleblowers, protestors, and leafleters.

Towards this end, attorneys from the Department recently met with representatives of the Humane Society and the ACLU to listen to their concerns and discuss the need for the proposed amendment. We explained how the law could not be used against those who mistakenly trespass during a lawful protest but only against those who commit unlawful economic damage or who ***intentionally*** instill a reasonable fear of death or serious bodily injury by a ***course of conduct*** involving threats, acts of vandalism, property damage, trespass, harassment, or intimidation.[1] Moreover, this law breaks no new legal ground because such intentional crimes are proscribed in the different context of domestic violence and stalking statutes.[2] Also, the law protects only those who have a ***reasonable*** fear of death or serious bodily injury and not the rare timid or fragile employee who might subjectively be frightened by legitimate protestors.

This proposed law builds on existing concepts in the federal criminal code. It is narrowly drafted to criminalize only outrageous, violent conduct that causes economic damage or instills the reasonable fear of death or serious bodily injury in others. Nevertheless, the Department has heard the concerns of the Humane Society and the ACLU and is willing to work with this Subcommittee to leave no doubt that nothing in this section shall be construed to prohibit any expressive conduct (including peaceful picketing or other peaceful demonstration) protected by the First Amendment.

The First Amendment is not, however, a license for the use or threatened use of violence, or for the commission of other crimes. Even if these crimes are politically motivated – even if they are committed as a form of protest – Congress is empowered to prohibit the conduct it deems offensive without offending the First Amendment. Those who cross the line from free speech to criminal conduct should be prosecuted and, if convicted, should be punished appropriately. As it has done in other contexts, the Congress should give prosecutors the tools to do so effectively.

Conclusion

Prior Congressional action has provided law enforcement and prosecutors with a solid framework within which to pursue the goal of prevention and disruption of violent extremism within our borders. We, as prosecutors in the Justice Department, have more work to do to eliminate this violent threat, and we urge you in Congress to continue to build upon and enhance the legal tools needed to accomplish our mutual goals.

Mr. Chairman, thank you again for inviting us here and providing us the opportunity to discuss how the statutes are being used around the country, consistent with our Constitutional values, to fight violent extremism through reliance upon the criminal justice system. We would

---

[1] The Humane Society would like to amend the bill to require the prosecutor to prove that any defendant "intentionally place[d] a person in reasonable fear of the death of, or serious bodily injury to that person . . . by a course of conduct involving . . . ***intentional*** trespass." To require a prosecutor to prove that a defendant intentionally did something intentional is redundant and inartful. The Department cannot envision any fact pattern where a putative defendant should be protected because he intentionally placed a person in reasonable fear of death or serious bodily injury by a course of conduct involving unintentional trespass.

[2] *U.S. v. Bowker*, 372 F.3d 365, 378-83 (6th Cir. 2004)

also like to thank this Committee for its continued leadership and support. Together, we will continue our efforts to secure justice and defeat those who would harm this country.

Mr. COBLE. The pressure is on you, Dr. Basso. Good to have you with us, Dr. Basso.

TESTIMONY OF MICHELE BASSO, Ph.D., ASSISTANT PROFESSOR, DEPARTMENT OF PHYSIOLOGY, UNIVERSITY OF WISCONSIN

Dr. BASSO. Thank you, Mr. Chairman, Ranking Member Scott, and other Congresspersons for the invitation to speak to you about my experiences.

I am an assistant professor of physiology at the University of Wisconsin. I am also an affiliate of the Wisconsin Regional Primate Center because of my work with nonhuman primates, and our goal is to try to understand the brain mechanisms of Parkinson's disease, which, as you know, is a debilitating disorder of movement.

We work together very closely with neurologists and neurosurgeons who develop state-of-the-art techniques for treating Parkinson's disease, and our goal is to understand how these techniques work and how to improve them in order to increase the quality of life for patients who suffer from movement disorders such as Parkinson's disease.

My experience with animal rights activity began about 3 years or so ago. And two organizations at the University of Wisconsin tried to purchase property immediately adjacent to two of the primate centers located on campus, the Harlow Lab and the Wisconsin Regional Primate Center. They rented it with an option to purchase, and their mission or their goal, stated goal, was to establish a holocaust museum for the monkeys that were killed in the research programs going on at the Primate Center.

The second thing was these two groups also started a meeting on campus monthly called Primate Vivisection, A to Z, where they talk about—or try to engage investigators to discuss animal research and also the use of animals in research. Because of the chancellor at the university's commitment to free speech, of course, these programs continue on campus, and they use university facilities.

Now, I first heard of these activists, the same groups, with respect to a protest that took place at the University of Wisconsin, and with targeting members of the University of Wisconsin at their homes. I was among eight of the faculty members and the academic staff who were targeted. Although they went to the wrong home—they didn't have my correct address—what they did was they appeared at homes with a truck that had a video monitor on it displaying images of animals in cages, and they shouted with bullhorns obscenities and defamatory statements about the persons in the home, went and rang the doorbell and ran away and various activity like that—activities such as that. They also handed out fliers with my photograph and contact information, as well as sort of defamatory statements regarding me and my research.

So in response to this—I was very nervous and concerned about my safety, so I tried to protect myself in two ways. The first was I removed my name from the Internet sites where you can go to the tax assessor's office and find out the property that a person owns by typing in their name. So I removed that from the Website. And the second thing I did was to hire an attorney to quit-claim deed my house into another name so that if someone were to go to the tax assessor's office, they would not be able to find out where I lived or my home address.

But within 2 months' time—less than 2 months, I started to receive magazine subscriptions. First they came slowly, but then they came rather aggressively. I have over 50-plus magazine subscriptions and various paraphernalia. I also received various books, the titles of which are things like fatal—"Oh, What a Slaughter," "Fatal Burn," "Predator," "The Perfect Orgasm," things like this. At the same time, I received two phone calls, voice messages, anonymous voice messages, through a messaging service that said something to the effect of, Hello, Michele, we know you're a monkey killer, and you can't get away from us. We hope you enjoy the magazines that you are receiving. And you will never get away, even though you tried to change the name on your house, things like that. So there were at least two of those messages.

So I guess I can't stress the critical impact that this has had on me and my ability to do my work. And I know that a number of my colleagues across the country experience similar targeting as well as more violent and aggressive—one colleague has had their house windows broken and their yards destroyed in California, for example.

So it's critical, also, to point out that the work that I do is subject to very strict regulations and oversight, and we have at least five animal care and use committees on campus that regulate what we do. And we also abide by the 3R principles for research: We reduce, refine or replace our animal models whenever possible. And when we are doing that already, we are required to justify why we don't do it even more. So working on animals, we believe, is a privilege, and one that we don't take lightly.

So I would like to just thank all of you for considering this important legislation and hearing my testimony. Thank you.

Mr. COBLE. I thank you, Doctor.

[The prepared statement of Dr. Basso follows:]

PREPARED STATEMENT OF MICHELE BASSO

My name is Michele Basso. I am an Assistant Professor of Physiology at the University of Wisconsin Madison Medical School. I am also affiliated with the Wisconsin Regional Primate Center because my laboratory studies primates as a model to understand Parkinson's Disease. Our research, which is funded by the National Institutes of Health and the Parkinson Disease Foundation, focuses on how the brain integrates visual information to produce movement. As you all know, Parkinson's disease is a debilitating neurodegenerative disease that is caused by a defect in the brain's ability to correctly initiate and control movement. I work together with neurologists and neurosurgeons across the country who treat Parkinson's patients with state of the art surgical therapies. In the laboratory, we use non human primates to understand the mechanisms of action of these therapies in order to improve them. Finally, our work on non human primates together with our work on humans will improve the quality of life of patients suffering from movement disorders such as Parkinson's disease.

About 3 years ago, an FBI agent opened a case for me because I received an email from an unknown source calling me an animal killer and equating me with Dr. Mengele, the war criminal who tortured humans during the Nazi era. At the same time, a colleague told me that my home address was circulating through an animal rights chat group. These events followed a Freedom of Information Act request for my animal use protocol, to the Director of the WI Primate Center.

Since then, animal rights activists have been active on campus. First, the Wisconsin Alliance for Animals and the Primate Freedom Group rented, with an option to purchase, a piece of property immediately adjacent to the two primate centers—The Harlow Lab and the WRPC. The groups referred to the building as the upcoming "Holocaust Museum" designed to remember the monkeys killed by researchers at the two primate centers. Second, some time around September 2005, a monthly meeting, held on the University campus called "Primate Vivisection from A to Z" began. Since the University is committed to the free speech rights of all members of the community, the Wisconsin Alliance for Animals and the Primate Freedom Group were granted access to University property and facilities for these meetings. The stated purpose of these meetings is to inform the University and surrounding community about the research going on at the Primate Centers and to establish a dialogue with investigators regarding the use of animals in research. These meetings are still occurring.

In October 2005, the UW Madison police department contacted me and informed me of an upcoming animal rights protest. I would be one of 8 University members who would be targeted at their homes.

The animal rights protest consisted of a truck with a video screen on three sides. The screens displayed images of non human primates in cages. The truck was parked outside people's homes and a group of activists with bullhorns harassed the people inside the homes. The name of the person was shouted along with accusations such as monkey killer or animal abuser. The protestors would run to the front door, ring the bell and run away. They circulated flyers about the individual to the neighbors. Over the course of one week, they attended the homes of 7 of the 8 people, the two primate center directors, one academic staff and 4 scientists. 5 of these 7 people directly targeted were female. Of the 51 non-human primate investigators on campus, only 11 of these are women.

The activists attempted to go to my home but made a mistake and protested in front of the wrong house. They circulated a flyer throughout the neighborhood containing my photograph and incorrect contact information as well as a number of misstatements regarding my research program and personal attacks on my competency. They also wrote with chalk on the sidewalk covering an area approximately 3 feet by 5 feet that said, "Basso Animal Abuser".

When people disagree they are entitled to exert their first amendment rights. For example, if a group does not agree with a potential legislative action, they protest at the government office or in public squares, but not in front of private homes. Protests at private homes serve what purpose other than to malign people and their children, intimidate and frighten families in their homes?

After this disturbing set of events, I attempted to protect myself by doing two things. First, anyone can go online to the tax assessor's office web page and look up a name to find a home address. I contacted the office and requested that they remove my name from the web site. Second, removal from the web does not eliminate access to the information. Anyone can still go to the office and look up personal information. So I hired an attorney to quit claim deed my home into another name. In this way, my name would not be associated with any property in Madison.

In slightly less than 2 months time, I received a magazine to which I did not subscribe. Then I received a couple more magazines. I started to receive statements from magazine companies indicating that I placed gift subscriptions to others on campus. I received in total approximately 50+ magazine subscriptions and other mail-order paraphernalia.

At the same time I received two anonymous voice messages from a messaging agency. Both messages had very similar content and I paraphrase: 'Hello Michele, we know you are a monkey killer. We hope you are enjoying the magazines you have been receiving. You cannot get away from us.' The second message said the same but included a statement like, 'you cannot hide from us even though you changed the name on your house. You will never get away from us.'

In addition to the magazine subscriptions, I received two book club subscriptions. Each arrived with an initial shipment of ~ hardcover books. Some of the titles of these books include "Fatal Burn" "Oh What a Slaughter" "Predator" "The Perfect Orgasm" and the like. As I am sure you can appreciate, these activities take up an enormous amount of my time. I was reported to a credit agency due to delinquency for a magazine subscription but because the FBI is investigating these events, I have a case number I supply to the companies to correct these issues.

It is critical to point out that biomedical research is subject to very strict regulations and oversight. We have an animal care and use committee for each school at Madison and an all campus committee that oversees all schools. My research meets or exceeds all standards set by the USDA, Public Health Service Policy as well as local guidelines for the care and use of non human primates in research. We abide by the well-known 3R principle concerning the use of animals. Whenever we can, we reduce the numbers of animals used, we replace the animal model with some other or we refine the technique we use to ensure maximal well-being of the animals. When we already meet the 3R requirements, we are required to justify why we cannot reduce or refine more. Working on animals is a privilege that neither I, nor my colleagues take lightly.

I would like to thank all of you for considering this important legislation. I believe it is important that we protect the free speech rights of all individuals. It is equally important for me to be able to come and go from work and my home and not feel threatened, intimidated, harassed or slandered. I have a right to live free of fear. Thank you very much.

Mr. COBLE. Mr. Trundley.

TESTIMONY OF WILLIAM TRUNDLEY, VICE PRESIDENT, GLOBAL CORPORATE SECURITY AND INVESTIGATIONS, GLAXOSMITHKLINE

Mr. TRUNDLEY. Good morning, Chairman Coble, Ranking Member Scott and Members of the Subcommittee. My name is Bill Trundley, and I'm the vice president of corporate security and investigations for GlaxoSmithKline.

GSK is targeted by animal rights extremists because of our relationship with Huntingdon Life Sciences. This is secondary targeting. Tertiary targeting involves similar violent attacks against companies and individuals merely because they have a relationship with GSK.

In the past 21 months in the U.S., GSK has experienced 150 incidents, including 75 intimidating home demonstrations and 10 cases of serious damage to property, which have traumatized employees, their families and their children. In all of these, the common thread has been fear, intimidation and the threat of violence, as you can see from this poster, which depicts a T-shirt with a picture of an AK-47 assault rifle and glorifying violence against researchers.

This other exhibit is a SHAC terror card used by extremists to threaten employees and their families. Some of these terrorist acts against our employees include mail theft, which reveal the spouse's alcohol treatment program. They left a bottle of beer at her door with a note saying, "Have a drink, bitch," and then went to her son's school to hand out this disgraceful flier to his classmates. They also left a message on the family phone saying, "We've been watching you, and we know you're alone."

Several employees' homes have been attacked at night, smashing windows while they slept. In Philadelphia, an employee was threatened by an extremist who yelled at her, "I have your license plate; we'll track you down and we'll kill your family." In Baltimore, an employee was contacted late at night asking her to come to the city morgue to identify a relative who died. On arrival, she learned the call was a hoax. One employee's 8-year-old son was so traumatized by these incidents, he would wake up at night staring out of the window, so scared that the terrorists would return. And in some cases, over 100 extremists have terrified employees at their homes like a baying, screaming lynch mob.

We have noticed an increase in the frequency and severity of these acts, which also involve others merely because of their association with GSK, and these include attacks against people who work for universities, charities and other companies. Their homes, cars and other property have been wrecked. And in one case a retiree in Long Beach, New Jersey, had his home and car damaged simply because his name was the same as a GSK employee's; he had no connection with GSK, and it was a case of mistaken identity.

Now, GSK has received excellent support from law enforcement, but continues to be targeted; and to date, none of the acts against GSK has resulted in a criminal conviction. This is because the current laws are inadequate. We believe that H.R. 4239 will enable law enforcement to deal effectively with these crimes, and we urge Congress to pass this legislation.

The situation today in the U.S. Is similar to what we experienced in the U.K. 5 years ago. In the first 6 months of 2001, GSK employees in London experienced over 3,000 separate actions by animal extremists. Prominent U.S. Extremists spent a year or more in the U.K. During 2002 where they were seen to associate with leading U.K. Extremists. At that time there was little support from the U.K. Police, who could never apply appropriate resources to deal with the situation.

The extremists became emboldened and placed fire bombs at the homes of our employees while their children were asleep in the house; destroyed entire buildings and other property; terrorized employees, their families and their children. A year ago, the U.K. Government introduced an effective piece of legislation and gave additional resources to the police. We are now seeing the benefits of this.

My advice to the Chairman and Members today is if the U.S. doesn't act now, they will face the same level of violent escalation and endangerment to the lives of American citizens.

I would like to finish by saying that before any new medicine can be used on humans, it is necessary to test their safety on animals, as required by Federal law. And those involved in scientific research are regular people trying to earn a living, raise a family and provide a decent future for their children. They are committed to the discovery of new medicines to help cure serious illness and disease; yet it is they and their families and associates who are left to suffer at the hands of violent extremists.

Thank you for the opportunity to appear before you today.

Mr. COBLE. Thank you, Mr. Trundley.

[The prepared statement of Mr. Trundley follows:]

PREPARED STATEMENT OF WILLIAM TRUNDLEY

Good afternoon, Chairman Coble, Ranking Member Scott and members of the Subcommittee. My name is Bill Trundley, I am the Vice President of Corporate Security & Investigations for GlaxoSmithKline and I want to thank you for the opportunity to address the Subcommittee regarding the impact of Animal Rights Extremists (ARE) on GSK employees as well as other individuals who have been targeted solely because of their relationship with GSK.

GlaxoSmithKline is one of the world's leading healthcare companies that discovers, develops, manufactures and markets pharmaceuticals, vaccines, over-the-counter medicines and health-related consumer products. Part of this work involves testing new medicines on animals to assess the safety of the substances before they are administered to humans, as required by law in just about every country in the world.

GlaxoSmithKline strongly supports and encourages passage of H.R. 4239 and its Senate counterpart to give law enforcement personnel the tools necessary to prosecute illegal animal rights activity.

Although GSK is a global research based pharmaceutical company, we are targeted by animal rights extremists because of our business relationship with Huntingdon Life Sciences (HLS). This tactic is referred to as secondary targeting because it attacks a company's customers rather than the company itself. In this case,

GSK is targeted because it has a business relationship with HLS. Even more outrageous is 'tertiary targeting' which seeks to punish businesses and individuals merely because they have some remote relationship with GSK or a GSK employee. Animal rights extremists have found these tactics to be effective because they exploit current law's inadequacy of addressing and protecting non-primary targets. HR 4239 will address this secondary and tertiary targeting and close the existing loophole that has been exploited in order to terrorize completely innocent people and legitimate businesses.

In the past 21 months in the US, GSK has experienced approximately 150 AR incidents directed at our employees, Board members, and others with only a tenuous connection to GSK. Unfortunately this list includes several non-profit organizations such as the Eisenhower Fellowship, universities, and private high schools.

Since January 2005, GSK employees and our friends and family have been subjected to approximately 75 intimidating and threatening home demonstrations and 10 cases of criminal damage to property in the United States. These incidents have traumatized employees and family members particularly children. In many cases, intimidating and defamatory flyers have been circulated to neighbors and classmates of children of GSK employees.

In all of these incidents the common thread has been fear, intimidation and the threat of criminal action, and in many instances the ensuing criminal acts demonstrate that these were not idle threats or mere free speech. *[Show exhibit at this point]* The exhibit entitled "SHAC Terror Card" is a typical flyer used at these demonstrations. This card proclaims "Do you do business with Huntingdon Life Sciences? . . . If you do, there's something you should know . . . Radical animal rights activists have been targeting executives and employees of companies that work with HLS" with criminal activity including: smashed windows; spray painted houses; glued locks; vandalized cars; stolen credit card numbers; ID theft; fraud; and continuous acts of harassment and intimidation against employees, their children and spouses. The card states that "the only way to end or prevent such attacks . . . is to stop doing business with Huntingdon." It is no coincidence that many of the threatened criminal acts in this flyer have been carried out against our employees and associates.

Some of the acts committed by those representing animal rights groups include:

- Theft of mail from a GSK employee, which revealed divorce proceedings and an alcohol treatment program recently completed by his spouse. Animal Rights Extremists left a bottle of beer at her front door and a note stating "Have a drink Bitch". The same day AREs visited the school of her son placing slanderous flyers throughout the campus depicting one parent as an animal killer and the other an alcoholic. Similar defamatory statements were e-mailed to the school's staff. *[Redacted version of Flyer to be show during this part]* On a previous visit to the spouse's residence an anonymous message was left on her answering machine stating "We have been watching you and we know you are alone."
- A GSK senior executive had his home attacked twice in the middle of the night resulting in spray painting of the exterior of the house with the words "Puppy Killer Dave" and a rock thrown through a large front window. He has also been subjected to anonymous late night threatening calls and numerous daytime intimidating demonstrations, where defamatory flyers and the SHAC "terror card" were distributed to neighbors.
- During a Hugs for Puppies (a NJ/PA based animal extremist group) protest at GSK's Philadelphia parking facility, a female GSK employee was threatened by a Hugs for Puppies protestor, when he yelled at her, "I have your license plate, we will track you down and kill your family."
- A GSK physician was contacted in the middle of the night by someone posing as an employee of the Baltimore City Morgue, requesting her to come to the morgue to identify a relative who had died. Upon arrival at the morgue she learned that the call was a hoax, and was then fearful that someone was lying in wait for her upon returning to her home in the middle of the night.
- Another GSK employee was subjected to several ARE demonstrations at his home, including leafleting the neighborhood with the SHAC Terror Card. The employee's eight year son was traumatized by the incident, waking up in the middle of the night staring out the window for fear that the terrorists would return.

Obviously GSK is very concerned about the targeting of its employees and we've noticed an upsetting trend in the frequency and increasing severity of these acts. While we will continue to protect our employees in an appropriate manner, it is worth mentioning that other individuals have been targeted merely because of their association with GlaxoSmithKline. For instance:

- The President of a New York University had his house spray painted because he invited someone to speak at the mid-year commencement address who happened to sit on the same board of another organization with a GSK employee.
- An employee of the Eisenhower Fellowship had acid or paint stripper thrown onto two vehicles parked at her residence just beneath an open window near where her young son lay sleeping. The employee at the time was expecting her second child. Eisenhower is a non-profit organization who happens to have a GSK executive on their Board.
- Eisenhower Fellowship also had the locks of their building glued causing the expense of replacing the damaged locks.
- A Long Beach Island, NJ, retiree had his home and car spray painted simply because he had a name similar to a GSK executive. He has no affiliation with GSK or HLS. Animal Rights Extremists have been arrested for this crime and are awaiting trial.
- A Philadelphia area executive serving on a Board of Directors with a GSK Senior executive was subjected to character assassination solely because of the GSK executive presence on the same board. After having his mail stolen from his residential mailbox, country club members were informed he was a pedophile by a forged letter purportedly from a fellow club member. An invitation to an anniversary dinner was also stolen, resulting in an obscenity laden message to the hostess threatening intimidation if the individual wasn't uninvited from the private dinner party.
- A senior executive of a Fortune Five Hundred specialty chemical company had his home spray painted and his car doused with acid or paint stripper, again solely because a GSK executive serves on their Board.

GSK has received excellent support from law enforcement, and is appreciative of the efforts by agencies such as the FBI, Philadelphia Police Department, and other State and Local law enforcement agencies. Despite this support, GSK continues to be targeted with intimidation and criminal acts, and to date none of the acts against GSK has resulted in a criminal conviction, despite the tireless efforts of law enforcement. We believe this is because the existing laws are inadequate to provide law enforcement and prosecutors with the tools necessary to bring these terrorists to justice. GSK believes House Bill 4239, the Animal Enterprise Terrorism Act, will provide law enforcement with those tools, and we urge Congress to pass this legislation.

It is worth mentioning here that the situation today in the USA is very similar to what we experienced in the UK five years ago. I can tell you that in the first six months of 2001, GSK employees living and working in and around London experienced over 3,000 separate terrorist actions by animal extremists. I am aware that leading lights in the US animal extremism movement spent up to a year or more in the UK during 2002/3 where they were seen to be associating with the leading lights in the UK extremist groups. At that time, there was little or no support in the UK from the police who claimed that, as there was no resolve by government, they could not apply the appropriate resources to deal effectively with the situation. Because of this, the extremists became more emboldened and have placed firebombs at the homes of our employees while their children were asleep in the house, destroyed property, terrorized children and caused incredible stress on individuals and their families. A year ago, the UK government introduced more effective legislation and provided the police with the necessary resources. We are now seeing the benefits of these measures. Had it not been for the introduction of effective legislation, and its application by the police, the situation in the UK would have worsened and my advice to the Chairman and members today, based on personal experience, is that if the US doesn't act now, they will face the same level of escalation and similar acts of violence, intimidation and the endangerment of lives of American citizens. The patterns of offending and extremist behavior are the same.

I would like to finish by saying that those involved in scientific research are regular people, trying to earn a living, raise a family and provide a decent future for their children. They are committed to the discovery of new medicines to help cure serious illness and disease yet it is they and their families and associates who are left to suffer at the hands of violent extremists.

Thank you for the opportunity to appear before you today.

Mr. COBLE. Mr. Potter.

TESTIMONY OF WILLIAM POTTER, JOURNALIST

Mr. POTTER. Good morning, Chairman Coble, Ranking Member Scott, and Members of the Committee.

Mr. COBLE. Mr. Potter, if you would suspend a minute, I won't penalize you. I want to recognize the distinguished gentleman from Florida Mr. Feeney, and the distinguished gentleman from Ohio Mr. Chabot, who have joined us.

And you may continue, Mr. Potter.

Mr. POTTER. Thank you, sir.

Good morning, Chairman Coble, Ranking Member Scott and Members of the Committee. I am honored to be invited to discuss civil liberties concerns raised by the Animal Enterprise Terrorism Act. As should be clear from the outset, though, I'm not a lawyer, I'm not a first amendment scholar, and I am not a spokesperson for the animal rights movement or underground groups. I'm here because of my freelance reporting.

I've written for the Chicago Tribune, The Dallas Morning News, Legal Affairs and other publications. And since 2000, I've closely followed the animal rights and the environmental movements, and the corporate-led backlash against them. I've documented an increasingly disturbing trend of terrorist rhetoric, sweeping legislation, grand jury witch hunts, blacklists, and FBI harassment reminiscent of tactics used against Americans during the Red Scare. The Animal Enterprise Terrorism Act is a continuation of that trend. The bill is ostensibly a response to illegal actions in the name of animal rights.

Department of Justice officials have told this Congress that their hands are tied in prosecuting these crimes, but their press statements tying the arrests of so-called ecoterrorists don't match this rhetoric. Just this weekend, four individuals were indicted for the 1998 fire at a Vail ski resort. The Government recently rounded up over a dozen environmental activists in the Northwest for alleged property crimes, and six animal activists were convicted in March of animal enterprise terrorism and other charges.

If this Committee wants surveillance, round-ups and convictions of animal activists, that's already underway. Law enforcement has not proven the need for heavier-handed tactics. Property crimes are already punishable as so-called animal enterprise terrorism. This bill, though, further expands that sweeping category to include protests, boycotts, undercover investigations, whistleblowing and nonviolent civil disobedience. The bill criminalizes any activity against an animal enterprise or any company tangential to an animal enterprise that causes economic damage defined as including the loss of profits. That's not terrorism, that's effective activism.

Businesses exist to make a profit, and if activists want change, they have no choice but to tug at pursestrings. That principle guided the grape boycotts of the United Farm Workers, the lunch counter civil disobedience of civil rights activists and the divestment campaigns of antiapartheid groups. Those tactics all hurt profits, and those tactics, if directed at an animal enterprise, would all be considered terrorism under this legislation.

Exceptions were made in the bill for losses from public reaction to information about an enterprise, but that's not an adequate safeguard. Corporations could argue that undercover investigators and whistleblowers hurt profits beyond public reaction. Those activists may cause a financial loss because they received a salary or prompted extensive employee background checks or prompted additional security measures.

Perhaps the greatest danger of this legislation, though, is that it will impact all animal activists, even those that never have to enter a courtroom. The reckless use of the word "ecoterrorism" by corporations and the Government has already had a chilling effect, and this legislation will compound it. Through my reporting I've already heard the widespread fears of activists that they may soon be labled terrorists, even for legal activity. They point to media smear campaigns by industry groups like the Center for Consumer Freedom, and many were stunned by full-page anonymous adds in both *The New York Times* and *The Washington Post* with a figure in a black mask labeling animal rights activists as terrorists.

They are also keenly aware that the Department of Homeland Security does not list right-wing terrorists on the list of national security threats, as in the Congressional Quarterly article I brought today, but puts animal activists at the top of that list.

This legislation will add to this climate of fear and distrust, and it will force Americans to ask themselves, is it worth it? Is standing up for my beliefs really worth the risk of being labeled a terrorist? That is not a choice that anyone should have to make.

Other activists may soon be asking themselves the same questions though. Prolife groups have already raised concerns that this bill could become the model for liberals in a changed Congress to target antiabortion acts as terrorists.

Public fears of terrorism since the tragedy of September 11th should not be exploited to push a political agenda. I urge you to reject this bill and ensure that limited antiterrorism resources are used to protect national security and human life, not profits.

Thank you, again, for this opportunity, and I look forward to your questions.

[The prepared statement of Mr. Potter follows:]

PREPARED STATEMENT OF WILL POTTER

Good morning Chairman Coble, Ranking Member Scott, and members of the committee. I am honored to be invited to discuss civil liberties concerns raised by H.R. 4239, the Animal Enterprise Terrorism Act.

I should be clear from the outset, though. I am not a lawyer. I'm not a First Amendment scholar. And I'm not a spokesperson for the animal rights movement, or underground groups.

I'm here because of my freelance reporting. I have written for publications including *The Chicago Tribune, The Dallas Morning News,* and *Legal Affairs*. And since 2000, I have closely followed the animal rights and environmental movements, and the corporate-led backlash against them. I've documented an increasingly disturbing trend of "terrorist" rhetoric, sweeping legislation, grand jury witch hunts, blacklists, and FBI harassment reminiscent of tactics used against Americans during the Red Scare.

The Animal Enterprise Terrorism Act is a continuation of that trend.

The bill is ostensibly a response to illegal actions committed by underground activists in the name of animal rights. Business groups have lobbied for this legislation for years. And Department of Justice officials have said they need help prosecuting these crimes.

At the same time, they have been patting themselves on the back for arresting so-called "eco-terrorists." Just this weekend, four individuals were indicted for the 1998 fire at a Vail ski resort. Earlier this year, the government rounded up over a dozen environmental activists in the Northwest for property crimes. And on top of that, six animal activists were convicted in March of "animal enterprise terrorism" and other charges.

If committee members want law enforcement to focus resources on the animal rights and environmental movements, that's already being done. The government has been able to make arrests and convictions using existing laws.

This legislation will not help solve crimes. It will, however, risk painting legal activity and non-violent civil disobedience with the same broad brush as illegal activists. It takes the administration's "you're either with us or against us" mentality of the War on Terrorism and applies it to activists.

This legislation criminalizes any activity against an animal enterprise, or any company connected to an animal enterprise, that causes "economic damage." That includes the replacement costs of lost or damaged property or records, the costs of repeating an interrupted or invalidated experiment, and "the loss of profits."

That clause, "loss of profits," would sweep in not only property crimes, but legal activity like protests, boycotts, investigations, media campaigning, and whistleblowing. It would also include campaigns of non-violent civil disobedience, like blocking entrances to a laboratory where controversial animal testing is taking place.

Those aren't acts of terrorism. They are effective activism. Businesses exist to make money, and if activists want to change a business practice, they must make that practice unprofitable. That principle guided the grape boycotts of the United Farm Workers, the lunch-counter civil disobedience of civil rights activists, and the divestment campaigns of anti-apartheid groups.

Those tactics all hurt profits. And those tactics, if directed at an animal enterprise, would all be considered "terrorism" under this bill. In fact, those three examples would probably receive stiffer penalties, because they caused "significant" or "major" economic damage or disruption. In other words, the more successful that activists are, the greater terrorist threat they become under this bill.

It is my understanding at the time of drafting this testimony that proposed changes might exclude "expressive conduct (including peaceful picketing or other peaceful demonstration) protected from legal prohibition by the First Amendment." It is a positive, yet incremental, first step to include peaceful picketing. However, the bill does not specifically exclude other activity like boycotts, whistleblowing, undercover investigation, and non-violent civil disobedience.

Furthermore, the inclusion of "trespassing" in damaging and disruptive activity puts undercover investigators and whistleblowers further at risk. Undercover video and photography undoubtedly impact profits. They have also led to prosecutions, animal welfare reforms, and a more informed democratic process on these issues.

Exceptions are made in the bill for disruption or damage "that results from lawful public, governmental, or business reaction to the disclosure of information about an animal enterprise." But this is no safeguard. For instance, undercover investigators and whistleblowers may cause financial loss for a company beyond the losses related to third party reactions. Companies may argue that salaries for undercover investigators, increased internal security, and extensive employee background checks are added costs of doing business because of activists. In short, this exemption seems to pose more questions than it answers.

You probably have noted that I have not focused on the clauses of this legislation dealing with significant bodily injury or death caused by activists. Those provisions are each problematic, but they are also, in some ways, non-issues. It's unlikely that even illegal, underground activists like the Animal Liberation Front would be impacted. Their actions, such as releasing mink from fur farms, spray-painting buildings, and arson, have not claimed a single human life.

This legislation will impact all animal activists, even if they never enter the courtroom. It will add to the chilling effect that already exists because of "eco-terrorism" rhetoric by corporations, lawmakers and law enforcement. Through my interviews with grassroots animal rights activists, national organizations, and their attorneys, I have heard widespread fears that the word "terrorist" could one day be turned against them, even though they use legal tactics.

They point to full-page anonymous ads in both *The New York Times* and *The Washington Post* this month, labeling animal rights activists "terrorists." The ads promote a website, *www.nysehostage.com,* that says "anti-business activists" like the Teamsters, Communication Workers of America and Greenpeace could be the next "eco-terrorists." Media campaigns by the Center for Consumer Freedom and other industry groups have used similar rhetoric to smear legal activist groups.

Activists also feel that the government is disproportionately focusing resources and attention on the animal rights and environmental movements. They cite reporting by *Congressional Quarterly* that showed the Department of Homeland Security does not list right-wing terrorists on a list of national security threats.

Those groups have been responsible for the Oklahoma City bombing, the Olympic Park bombing in Atlanta, violence against doctors, and admittedly creating weapons of mass destruction, but animal rights activists still top the domestic terrorist list.

This legislation will add to this fear and distrust, and will force Americans to decide if speaking up for animals is worth the risk of being labeled a "terrorist," either in the media or the courtroom. That's not a choice anyone should have to make.

Animal rights activists have been among the first victims of this terrorist scaremongering, but if it continues they will not be the last. Changes in the Supreme Court seem to have revitalized the anti-abortion movement, which, unlike the animal rights movement, has a documented history of bloodshed. But there's also a potential for backlash if upcoming elections alter the balance of power in Washington. Some anti-abortion organizations, like the Thomas More Society, have already raised concerns that this legislation could become a model for labeling other activists as terrorists.

All Americans should be concerned about this trend, regardless of how they feel about animal rights. The word terrorism should not be batted around against the enemy of the hour, to push a partisan political agenda. Public fears of terrorism since the tragedy of September 11th should not be exploited for political points. I urge you to reject this legislation in its entirety, and ensure that limited anti-terrorism resources are not spent targeting non-violent activism.

Thank you again for this opportunity, and I look forward to your questions.

Mr. COBLE. Thank you, Mr. Potter, and thanks to each of you panelists.

Now we impose the 5-minute rule against ourselves as well, folks, and I suspect we'll probably have a second round because this is a very significantly important issue.

Dr. Basso, I think you correctly stated that most—well, strike that—many—I remember you said most, but I'm sure many activists do so properly and lawfully. Unfortunately, they're tainted with the same brush with which those who don't do it lawfully. That's the unfortunate feature here. And it is my belief—now, I'm not suggesting these people are terrorists, they may be, but terrorists generally are cowards, or they conceal their face with black masks. They strike, as you pointed out, Doctor, at night, by dark of night. It bothers me.

Now, I'm not interested in trampling on anybody's first amendment rights. As I mentioned earlier, and the Ranking Member mentioned in his statement, we have a manager's amendment, but I really don't know that that was important because I think the bill, on its face, indicates first amendment protection.

But, Doctor, if you will, tell us in some detail how critical animals are in your research and its potential benefit for mankind—briefly if you can, because I'll need to get around to these other guys.

Dr. BASSO. The use of animals in research is critical. Virtually every major advance in the last century has depended upon the use of animals. In my research in particular, we coordinate very closely with clinicians, neurosurgeons and neurologists, and we try to ask as many questions as possible of the human brain, but we have to remember also those patients with whom we work are undergoing surgical procedures, and so they're there principally to be treated for their disease. What we then need to do is go back into the laboratory and replicate either the disease or the treatment in order to understand how it's working and how to make it better for the next time.

Mr. COBLE. I thank you.

Mr. McIntosh, some might indicate that the recent convictions of the six SHAC extremists to which you alluded in New Jersey might well suggest that no additional legislation such as this before us is necessary. What say you to that?

Mr. MCINTOSH. Thank you, Mr. Chairman.

We do not agree that the recent SHAC convictions indicate that no additional legislation is necessary. As I mentioned, we were forced to use the stalking statute there because, as has been indicated explicitly on a number of Websites, animal rights extremists explicitly attempt to tailor their tactics to avoid Federal jurisdiction, Federal investigation and prosecutorial jurisdiction. And so we believe that following successful prosecutions under the stalking statute, we would see another permutation to move to a situation where those extremists are trying to avoid the stalking statute perhaps by targeting not individuals, but entities.

Moreover, we think that clarity in the law on how broadly section 43 of the Animal Enterprise Protection Act applies is an independent good thing because it gives us a sense of what the actual crimes these extremists may be committing are, and thus doesn't force prosecutors to hunt through title 18 of the Code to find a crime they may have committed while attempting to avoid section 43. And second, it gives them a sense for what the Federal crimes actually are, what the scope of the Federal crimes are. And the scope of section 43 right now is not clear.

Mr. COBLE. I thank you, Mr. McIntosh.

Let me try to beat the red light by putting a question to Mr. Potter.

Mr. Potter, in your testimony you expressed concern that nonviolent civil disobedience would be criminalized under this bill. Let me ask you this, sir: Do you believe that spray painting abusive graffiti on people's homes and vandalizing homes and businesses or pouring acid on cars, do you think that is nonviolent?

Mr. POTTER. I think those are absolutely crimes, and they're absolutely not nonviolent civil disobedience.

Mr. COBLE. So you say that would be violent civil disobedience.

Mr. POTTER. I think that because I'm not an attorney——

Mr. COBLE. And I'm not trying to entrap you.

Mr. POTTER. I think they're absolutely crimes——

Mr. COBLE. Okay. I didn't understand you clearly.

My red light appears, and I'm just pleased to recognize the distinguished gentleman from Virginia.

Mr. SCOTT. Thank you, Mr. Chairman.

Mr. McIntosh, you've indicated that you prosecute some and don't prosecute others, and make good judgments along those lines. One of the concerns I have is what you prosecute and don't prosecute ought not be your discretion, but ought to be how the law is written. Let me ask a couple of questions along those lines.

What is the law for everybody else in other kinds of crimes, other kinds of protests in terms of business losses? How does the treatment in this circumstance differ from other kinds of protests?

Mr. MCINTOSH. Let me make an initial point here, which is that the loss of profits provision is not new to this bill, it's currently—it's part of the current law; so we are not proposing to change that provision. Section 43(d)(3) of the current law includes loss of profits in the current law.

It is my understanding that in this case we're seeing a specific targeting—we're seeing a set of tactics used in the animal enterprise terrorism or animal enterprise extremism front, and in the ecoextremism front more broadly, that is explicitly aimed at using violence against—violence and threats of violence against people to cause them to disassociate themselves, perhaps by imposing a loss of profits on—I'm sorry, to disassociate themselves with the animal enterprise. So this is a tactic we're seeing specifically in this area that we're not seeing in many other places.

Mr. SCOTT. Well, if it were to occur in another situation or another cause, why should that, too, not be illegal?

Mr. MCINTOSH. Congressman, I think that we would suggest that if some other cause adopted a similar set of tactics in an attempt to commit this sort of violence extortion, that similarly it ought to be illegal.

Mr. SCOTT. Are you not concerned that when you make these things cause-specific, you get into freedom of speech content?

Mr. MCINTOSH. Sir, we are not interested in what the cause is. If there are—if this is being used against other causes, we would be happy to see the ability to prosecute it as well. We are interested only in the tactics involved.

Congress has seen fit to pass an animal enterprise terrorism statute, and we are happy to prosecute it as it exists. And if there are proposals to give us that authority, a similar authority for other causes, we would be happy to prosecute those as well, sir.

Mr. SCOTT. The bill provides for the same treatment for conspiracy attempts as the completed offense. Is there precedence for that?

Mr. MCINTOSH. I believe there is, sir. In a number of instances they are treated the same. And I would be happy to, after the hearing today, get back to you with a list of similar places where—of places where they're treated similarly. We believe it is justified to do so because when two or more people conspire, they often can commit greater damage than an individual person. And moreover, they often lead themselves through sort of egging one another on to complete a conspiracy that an individual would not feel compelled to complete.

Mr. SCOTT. Well, that's—if you have the conspiracy or the attempt treated the same way as a completed offense, there's no incentive to discontinue.

Mr. MCINTOSH. Well, the attempt actually involves an attempt to succeed, so discontinuance is not a subject; that's a failed attempt to complete.

Conspiracy, on the other hand, is a situation where we think that often a conspiracy to commit the substantive offense, because of the greater damage it threatens, can be worse than an individual attempting to commit the offense on his or her own.

Mr. SCOTT. I yield back.

Mr. COBLE. The gentleman's time is expired.

The distinguished gentleman from Florida Mr. Feeney.

Mr. FEENEY. Thank you, Mr. Chairman.

Dr. Basso, first, do you believe that you were targeted because of the research you were doing to try to understand the causes and treatment of Parkinson's disease, and was that research funded by the National Institutes of Health, in part?

Dr. BASSO. Excuse me. I'm not sure why I was targeted, frankly. And yes, the research is funded by the National Institutes of Health and also the Parkinson's Disease Foundation.

Mr. FEENEY. Thank you.

Mr. Potter, I appreciate that you don't have a legal background. In your testimony you oppose the bill because you say that it—and I quote, "it criminalizes any activity," that causes economic damages. But the truth of the matter is the bill is very specific; it does not criminalize any activity, it criminalizes activity that either intentionally damages, disrupts or causes the loss of any property. So destruction or damaging property, or, alternatively, intentionally places in reasonable fear of the death of or serious bodily injury to a person, that's the traditional legal definition so that you understand it, that's called an assault. A battery is actually attacking somebody physically; an assault is the imminent threat to do so under traditional tort understanding. These are already illegal activities.

The point is that people are behaving illegally to make political purposes. In my view, you're just flat out wrong. They ought to be responsible for the natural and consequential damages of their disruptive behavior. There are first amendment protections that all of us believe are very important to this country, but I would advise you not to be making statements that any activity is criminalized because it's just flat out false. And maybe next time you'll want to consult—go ahead, you can answer.

Mr. POTTER. Well, Congressman, with all due respect, I'd like to point out that the definition given of economic damage means the replacement cost of lost or damaged property or records, the cost of repeating an interrupted——

Mr. FEENEY. Mr. Potter, we'll have to get you a logic course that you can understand one step to the next——

Mr. POTTER. If I can just finish. The easier is the loss of profits, and I think that's what would give any——

Mr. FEENEY. Reclaiming my time. I point out that the gentleman simply doesn't understand. You're not responsible for any of the definition you just talked about unless you have intentionally damaged or destroyed property or threatened somebody's life or bodily injury. So all of what you're referring to is not of concern if you behave legally. I want to assure you and advise you to go talk to an attorney before you come and testify before the United States Congress about what bills do when, in fact, they do not do.

If you commit a crime, then you may be responsible for some of those damages, and then the definition of what you're responsible for is important. But as long as you have not committed a crime, I want to assure you there's nothing in this bill that would make you a target of obligation for those economic damages.

Mr. McIntosh, I do believe that the gentleman from Virginia raises an important point, because whether or not you're trying to protect animals, or whether or not you're trying to protect—whatever issue you have, ultimately the goal is to protect a monkey or an unborn life or whatever issue you may have, and it is a concern that, as opposed to attacking the act when the act is the spray paint or the act is the imminent threat, I mean, it is a concern of mine that we are identifying specific causes, as worthwhile as they may be, for specific crimes. And you indicated that you're more concerned about the act than the goal as well. Is it fair to say, does the Justice Department itself take a position on that?

Mr. MCINTOSH. That is correct, Congressman. We are more—we are apolitical in this. We have no interest in the cause in question, we have only the interest of ensuring that the tactics used to advance that cause are lawful. It is our intention to prosecute unlawful acts without regard to the cause of——

Mr. FEENEY. One concern that I have in your testimony, you suggest that—and of course you haven't said this is criminalized by the act—but on page 3 of your testimony you said that one of the economic activities that causes—well, one of the activities that political groups use is Internet posting of home telephone numbers of law-abiding employees. I'm not aware of any Federal or State statutes that they may violate. If I post on the Internet my neighbor's address or telephone, is that a Federal crime?

Mr. MCINTOSH. Sir, that is not a Federal crime. The Federal crime is if you were to post that information in connection with a threat of violence that would put a reasonable person in fear for harm or death to himself or someone else——

Mr. FEENEY. So it's attached to the assault definition, genuine imminent concern about an attack.

Mr. MCINTOSH. That's right. This is what the courts call a true threat, where you post a person's name along with that——

Mr. FEENEY. Well, maybe in future testimony you will make it clear that you're not concerned about just mere posting of addresses and telephone numbers, it's combined with the other threat aspects that concern you.

Mr. MCINTOSH. Congressman, to the extent I didn't make that clear, I apologize.

Mr. COBLE. I thank the gentleman from Florida.

The distinguished gentleman from Massachusetts, Mr. Delahunt.

Mr. DELAHUNT. Thank you, Mr. Chairman.

Dr. Basso, your experience clearly is unfortunate and unacceptable. I think we all concur that it should not be tolerated. But I am going to direct my comments to the Deputy Attorney General.

You know, as I see and read the various cases that are reflected in the memorandum, I don't see a single case that would not fall within the purview of multiple—multiple State statutes, as well as a variety of Federal existing statutes, not necessarily just simply focused on the so-called Animal Terrorist Enterprise Act.

You know, the former Attorney General under President Reagan, Ed Meese, expressed his concern about the federalization of crime in this country, and, to be candid with you, I think that this could very well serve as an example. You know, you had to go to the—I guess the stalking statute to effect the indictments of those who were responsible—purportedly, allegedly—for the burning of a building out in Vail, Colorado; but I'm not convinced that there are not sufficient tools already to deal with the cases that are illustrated in the memorandum, as well as related by Mr. Trundley and Dr. Basso. I mean, as the gentleman indicated an assault, what about civil rights actions, both at the State and the Federal level?

You know, if there's a conspiracy here, if there's an organized enterprise, why not invoke RICO? Respond if you would.

Mr. McIntosh. Sir, I would be happy to do so. Let me first state that the Vail indictments actually involve an arson indictment. So they are not under section 43.

Mr. Delahunt. But that's my point. You're making my case for me. And I'm sure under a State statute arson carries a significant sanction.

Mr. McIntosh. Let me try and address this question with regard to the way these cases are investigated and prosecuted. In many cases these are not just local actions, we're talking about nationwide, and indeed in SHAC's case, an international conspiracy that——

Mr. Delahunt. I understand that because I just perused the memorandum. But in a RICO investigation, for example, why not utilize the RICO statute?

Mr. McIntosh. We have a——

Mr. Delahunt. I mean, you have predicate crimes.

Mr. McIntosh. In many cases SHAC has, and other animal extremists have tailored their crimes specifically, their campaign specifically to avoid committing predicate acts so that we can use RICO. Recently, the Supreme Court ruled in the Schindler case that a Hobbs Act predicate, which was a traditional hook for prosecuting these sorts of things, the traditional RICO predicate, was not available unless the defendant had gained for himself something of value. So if——

Mr. Delahunt. Well, again, you know, and I don't have much time, but do you work with local and State authorities during the course of the investigation of these crimes?

Mr. McIntosh. We absolutely do, sir.

Mr. Delahunt. I would dare say that, in many cases, that the sanctions that would be available under State statutes would be far more severe than the possible sanction under the Federal statute.

Mr. McIntosh. As a matter of law——

Mr. Delahunt. We're talking threats, we're talking assault, we're talking violation of civil rights. You know and I know that for each and every single conviction there's the possibility of a parole after sentence.

Mr. McIntosh. In many cases the——

Mr. Delahunt. Consecutive sentences.

Mr. McIntosh. We are seeing explicit attempts to commit low level harassment that in the end convinces people, despite the fact that these are low level actions of State crimes, that they are in a reasonable fear of death. And these are often crimes that have very low penalties individually when you look across the broad——

Mr. Delahunt. I will tell you, I can't—that's difficult for me. And I'm using Dr. Basso's case. An aggressive prosecutor in an investigation that would establish the violation of multiple State statutes, and given the parameters that you describe in terms of what this organization is about, would warrant, presumably, after conviction parole after sentences with considerable incarcerations. Now maybe that hasn't happened, but that's the problem of, you know, not looking—that's the problem in terms of recommendation to a court post jury verdict.

Mr. MCINTOSH. In many cases the incident against Dr. Basso cannot be connected by a local law enforcement agency to the action 2,000 miles away against someone else; whereas with a Federal offense we can connect those things and see them as a crime in both places committed by the same person.

Mr. DELAHUNT. If the Chair would indulge me for an additional minute.

Mr. COBLE. Just for 1 minute.

Mr. DELAHUNT. My point is on sentencing in State courts, the availability of bringing additional information before the court for sentencing purposes would clearly be allowed.

Mr. MCINTOSH. We are not seeing success with that in State and local law enforcement.

Mr. DELAHUNT. Well, I'm surprised. Have you consulted with the National District Attorneys Association on these cases?

Mr. MCINTOSH. I don't know that we have, but I would assume that we have and I am happy to get back to you with that information.

Mr. DELAHUNT. Well, again, my reluctance to embrace this is based upon a continuing federalization of State crimes. I understand there's a current statute on the books now, and it's not in any way an endorsement of the behavior that has victimized Dr. Basso and presumably others, but at some point in time we have to, you know, either respect the concept of federalism where these kind of crimes traditionally in our jurisprudence fall within the purview of the States. And if the States are not enforcing them, then it's a question of education, and insisting that State prosecutors work with Federal authorities in those cases where it's clear that this is an organized effort directed against Dr. Basso. And I just can't imagine, I can't imagine a State prosecutor not seeking the kind of penalties that would exceed whatever exists under Federal statutes.

Mr. COBLE. The gentleman's time is expired.

The distinguished gentleman from Ohio, Mr. Chabot, is recognized for 5 minutes.

Mr. CHABOT. Thank you, Mr. Chairman.

Mr. McIntosh and Mr. Potter, if I could begin with the two of you. How do members of the groups that we have been discussing here this morning communicate with each other in order to organize and plan their targeting campaigns? And perhaps, Mr. McIntosh, we could begin with you.

Mr. MCINTOSH. Congressman, we see the use primarily of the Internet to organize these campaigns. These tend to be often loosely affiliated groups that post the names and other biographical data of individuals online. And sometimes these are individuals associated with the actual enterprise, sometimes they're people associated with groups affiliated with the enterprise. I know that in Mr. Trundley's written testimony he also talks about targeting groups that are associated with, are entities associated with a group.

So these are a broad set of biographical data that are posted, often named as targets, and then they will list a set of tactics that ought to be taken against these people. And then when those tactics are taken by some anonymous entity, they are immediately, the next morning, up on the website saying some party did this, as we said they ought to.

Mr. CHABOT. Thank you.

Mr. Potter, anything you'd like to add to that, or could add to it?

Mr. POTTER. Yes. Thank you, Congressman.

I would like to add that when we were discussing first amendment speech, and even very controversial first amendment speech, like posting news of an illegal action, the Supreme Court has been extremely protective of first amendment activity, even in the most controversial——

Mr. CHABOT. Well, if you could just answer the question if you would. Do you know how they communicate with each other?

Mr. POTTER. From my understanding, it is through telephone calls, e-mail, the same way everyone communicates.

Mr. CHABOT. Okay. Thank you very much.

Dr. Basso—and I'm pronouncing that correctly, aren't I? Okay, thank you. Do you think that young scientists might seek other fields to pursue if intimidation and harassment continues on campus and at people's homes?

Dr. BASSO. Yes, I most absolutely do. I myself have considered leaving the field in light of my experiences. I have colleagues across the country who have experienced what I have experienced, and worse, and they have told me that they would leave. I have colleagues in the United Kingdom who are looking to leave because they feel they can no longer do their research.

So my concern in the long run is not only that young scientists won't go into the field, but already established scientists might even leave and go where the environment is more conducive.

Mr. CHABOT. Can you estimate how much of a financial burdenthat it's been for the University of Wisconsin to add extra security to protect those involved with research?

Dr. BASSO. Right. So I think that there has been somewhat of a small response to deal with some of these issues; and in large part I think because the seriousness of the problems is not fully recognized, and it may be in part because of a lack of a Federal legislation. So I think it's been a little slow.

But for me personally, my laboratory has been outfitted with alarm systems. And I know that we're moving toward increasing access to animal barriers, animal facilities and so forth. Not to mention my own time that I spend engaged in these activities, preventing me from engaging in my research efforts.

Mr. CHABOT. Could you comment on what State and local guidelines are in place for the humane treatment of the animals which you use in your scientific research and what decision-making body exists to determine when these guidelines have been breached?

Dr. BASSO. Animal research, and in particular non-human primate research, is subject to very strict regulations and oversight. And I'm not an expert in this area, so I can't give you all the details, but I know what my laboratory is required to do.

At the University of Wisconsin we have five animal care and use committees, one of which is an all campus committee that overseas the activities across the entire university. The five campus committees are from each of the individual schools, the medical school, the agricultural school and so forth. Those committees are made up of a number of scientists, a lay person, veterinarians. And their obligation and charge is to read through protocols that are submitted by scientists in advance of the work being conducted, and they have to abide by the rules and regulations set by the USDA, the guidelines set by the Public Health Service Policy, and also our own local policy rules.

Mr. CHABOT. Thank you. Mr. Chairman, could I ask unanimous consent to ask Mr. Trundley one additional question?

Mr. COBLE. Without objection.

Mr. CHABOT. Thank you, Mr. Chairman.

Mr. Trundley, throughout your testimony you detailed numerous acts of intimidation and criminal activity. How much would you say that GSK spends each year on security and cleanup because of the animal rights extremists?

Mr. TRUNDLEY. Well, sir, we do keep that information, but I would be prepared to give that to you personally outside of this meeting, because from our experience we would give that answer and then by the end of the day such information would be posted on an extremist website and they would be crowing with glory. We would just be giving them a platform on which they could grandstand. But I would be prepared to give that information personally later.

Mr. CHABOT. Okay. That would be fine. Would you say it is significant?

Mr. TRUNDLEY. It is significant.

Mr. CHABOT. Thank you very much.

Thank you, Mr. Chairman.

Mr. COBLE. I thank the gentleman. As I said previously, I think this issue warrants a second round, so let me editorialize a minute, the gospel according to Coble, Mr. Scott.

Some folks have indicated that if this bill were enacted it would have a chilling effect upon the animal activist activities. I think an equally convincing argument could be that some of the illegal activities by some of the animal activists could have a more obvious chilling effect upon more legitimate animal research by law abiding citizens. I think that argument ought to be presented.

Mr. Trundley, your body language told me that you wanted to insert your oars into the water as Mr. Delahunt was examining Mr. McIntosh. Did you, or did I misread you?

Mr. TRUNDLEY. No, you didn't misread me, sir.

Mr. COBLE. But far away, because Bill's not here to hear this, but——

Mr. TRUNDLEY. But the point is H.R. 4239 will be designed to assist the police, provide them with the tools and the necessary measures to investigate crimes against secondary and tertiary targets, whereas existing law is designed to protect the primary target. And despite the fact that we have this existing law at State and local level, there has still been no convictions for the crimes committed against GSK people, 150 in the last 21 months.

We need this legislation to enable the police to become proactive in the way they conduct their investigations. These people are organized along terrorist cells, independently operating and using the Internet and e-mails in order to—clandestinely, clandestinely to make contact with each other. And then the result of their activities are clandestinely posted on websites that do not operate within the United States or the United Kingdom.

Mr. COBLE. I thank you, sir.

Dr. Basso, if Mr. Scott had received a telephone call like you did, I might say, well, that's just some screwball having a little fun, they're mere words, Bobby, don't be upset about it. I could say that. But if that call came to me, if I was the beneficiary of such a call, Mr. Scott might say the same thing. But when you're the beneficiary, it takes on an altogether different meaning. I think you can attest to that. And I don't mean to be speaking for you, Doctor, but as evidenced by your testimony, you were placed in fear, were you not? I know I'm leading the witness with that question.

Dr. BASSO. No, you're absolutely correct. I was very fearful for my well-being, for the well-being of the laboratory personnel and for my animals, in fact. Yeah, this is very important.

Mr. COBLE. Well, I reiterate—I don't mean to be speaking for every Member of this Subcommittee, but I know each one of them on both sides, and I don't think any one of us is interested in trampling on first amendment rights, but this is a very, very difficult issue it seems to me.

Mr. Trundley, let me revisit the SHAC terror card. Talk to me about that a minute in my time remaining.

Mr. TRUNDLEY. Well, as you can see, sir, the terrorist, as we would describe them, is wearing a balaclava or a ski mask to disguise their faces. These are issues. The threats are quite explicit, stop doing business with Huntingdon Life Sciences, stop experimenting on animals. These are left with the scientists and members of their families. These are the people that are involved in bringing new medicines to the public, medicines such as flu vaccines, cervical cancer vaccines, breast cancer treatments. Without those people working on those products, you won't get them there. And they are terrified, they are traumatized, they are debilitated when something like that is handed to them personally or sent to them through the mail post, or during what is on the surface a peaceful demonstration their neighbors are contacted in person and handed a terror card such as this, or they're told your neighbor, who works for Glaxo Smith Kline, is a pedophile, is a puppy killer, is a murderer. It's designed to create terror and fear widespread, not just on the individual concerned, but on his colleagues when he goes to the office the following day, or with his colleagues and wider throughout the research and medical communities.

If it was informed, reasoned, peaceful debate, we would welcome that. We like to hear the views of others and we like to give our views as well, but in an articulate, controlled and informed manner.

Our point is, why create fear and terror amongst a group of scientists or those involved in medical research, not only scientists, but sales representatives, admin assistants and executives of the company, why create that environment of fear of posting it to the Internet?

Mr. COBLE. Well, as you said, Mr. Trundley, designed to create fear; actually not only designed to create fear but delivering fear.

Mr. TRUNDLEY. It achieves their objective.

Mr. COBLE. In spades. I see the red light.

The distinguished gentleman from Virginia.

Mr. SCOTT. Thank you, Mr. Chairman.

Mr. McIntosh, you have a copy of the bill and the—what's called the discussion draft before you?

Mr. MCINTOSH. I do, sir.

Mr. SCOTT. Okay. Mr. Potter has indicated that the bill may proscribe what are legitimate protests. In fact, if you have a successful protest—on page 2, line 9, you define the offense of whoever travels Interstate 1 for the purpose of disrupting and intentionally, on line 15, disrupts, that constitutes the crime which would really be the result of—a bona fide result of a successful protest.

The discussion draft, however, makes a change in that, and the offense is defined as someone who travels for the purpose of damaging or disrupting, and in connection with such purpose, A, intentionally damages or causes the loss of property, or intentionally places a person in reasonable fear. Now intentionally damaging or causing the loss of property is already a crime, damaging somebody else's property. Placing someone—intentionally placing somebody in fear is already a crime, that's assault. Does that change—should that change fix the problem that Mr. Potter has articulated?

Mr. MCINTOSH. Congressman, as I stated initially, we do not believe that the draft of 4239, the introduced bill, is unconstitutional. However, to the extent that there are concerns that it would show first amendment activity, I think it's clear that the discussion draft that I've been shown would go a long way toward remedying those concerns.

Mr. SCOTT. Mr. Potter, have you seen the discussion draft?

Mr. POTTER. Yes, I have.

Mr. SCOTT. Does that address the concerns that you have articulated?

Mr. POTTER. No, it does not, sir.

Mr. SCOTT. Okay. Why doesn't it?

Mr. POTTER. The main changes I saw in the discussion draft were, at the end, the specific exclusion of activity like picketing or lawful demonstrations. I'd like to point out that we would hope that would already be included under our conception of protected activities. So to point it out almost implies and acknowledges the overly broad and vague language of this legislation and the true danger it poses to first amendment activity.

And furthermore, that language still does not prohibit the use of this animal enterprise terrorism clause against things like civil disobedience, and perhaps even whistleblowing and undercover investigations.

Mr. SCOTT. Well, in connection with that, you would have to intentionally damage or cause a loss of property, or intentionally place someone in reasonable fear of death or serious bodily injury.

Mr. POTTER. My concern is that language, "the loss of property," is extremely broad and vague. I know the Congressman mentioned that there are specific definitions at play here, but I think when an average person reads the loss of property, any activist will see that and say, that's what I do, you know, I'm trying to impact a loss of profit to influence their business decisions.

Mr. SCOTT. You're using property to include profits?

Mr. POTTER. Excuse me, I misspoke. I'm talking about profits. But this says intentionally damages or causes loss of property in the discussion draft, page 2, starting on line 1.

Mr. POTTER. And I'm also operating—I'm looking at page 5, 3a, when we're defining economic damage and disruption with the loss of profits as specifically laid out. That was part of the crux of my concern of incorporating not only first amendment activity, but also civil disobedience through that definition.

Mr. SCOTT. Okay. The economic damage comes in the penalty part. If you have caused—if you have violated the law and caused a loss or damage to property, then you are guilty; the penalty comes in for the economic damage. Maybe we need to review the cross references, but the fact that you have to be exposed under the law, you have to actually damage or cause the loss of property, which is a crime already.

Mr. POTTER. I'm sure decisions will be made about the interrelation of these definitions in those actual clauses.

In addition to that, my overriding concern is that, regardless of that minutia, using the word "terrorism" to go after things that are already crimes, such as property crimes, and also potentially, as I've raised, first amendment activity, will have an overwhelming, chilling effect and add to the chilling effect that's already going on by using the words ecoterrorism, animal enterprise terrorism in a post-9/11 climate.

Mr. SCOTT. Thank you.

Mr. Chairman, can I ask one additional question of Mr. Trundley?

Mr. COBLE. If you will suspend just a minute, Mr. Scott. Folks, we have a vote. I'm inclined—I hate to keep the panel here, but, Bill, how long will it take you?

Go ahead, Mr. Scott.

Mr. SCOTT. Have you sought civil remedies against the people who were involved in these crimes?

Mr. TRUNDLEY. In the United Kingdom we have, so, yes, but the opportunity has not presented itself within the United States.

Mr. SCOTT. Have those actions been successful?

Mr. TRUNDLEY. In the United Kingdom it was successful in that we managed to obtain what's known as a John Doe High Court Injunction; that is, an injunction threatening contempt of court against persons unknown, which is completely novel and new in the U.K.

Mr. SCOTT. But no civil damages?

Mr. TRUNDLEY. No, but once the individuals have been identified, we would serve that injunction upon them, and then go for damages as well.

Mr. SCOTT. Thank you, Mr. Chairman.

Mr. COBLE. I thank the gentleman. The gentleman from Massachusetts.

Mr. DELAHUNT. I will be very brief.

Mr. COBLE. Bill, if you will suspend, if you need more time we can adjourn and come back.

Mr. DELAHUNT. No, that will be unnecessary.

In terms of the line of questioning by Mr. Scott, I think the concerns regarding first amendment issues can be resolved by more closely drafted language. I am still at the point, however, where I am not convinced that we are going down a road of federalizing criminal law. And I know Dr. Basso wanted to say something in response to the questions I asked earlier, and so let me give her that opportunity.

Mr. COBLE. And Doctor, if you could be brief, we won't keep you all here, but we have to go vote, so if you could be terse.

Mr. DELAHUNT. This will wrap it up anyhow.

Dr. BASSO. Sure. I guess I was thinking as you were talking that there's a particular organization that is moving from State to State and establishing little niches of animal rights activity across the country, and it struck me that that's more an important issue that we as a nation should be taking into account rather than a State to State——

Mr. DELAHUNT. But my point, Dr. Basso, is that clearly, for example, the Post is showing the AK-47. Most States have State statutes which prohibit threats——

Dr. BASSO. Right.

Mr. DELAHUNT. You know, I would suggest that that State statute would just clearly have been violated with that particular depiction because the most reasonable inference would be that it was intended to create fear, to threat, to provide a threat. And the point is that State prosecutors will often act much more quickly and expeditiously where aware or fully informed of the concerns that both you and Mr. Trundley, I believe, have provided us, have testified to, that there is more than enough criminal sanctions out there today. It's a question of, in my judgment, educating law enforcement both at the State and local, as well as the Federal level, because Federal resources are very limited. It is only, I dare say, those high profile cases; for example, out in Vail, Colorado, it was a cause celebre, if you will out there, that directed the attention of the Federal authorities to pursue it.

So in any event, you and your associations, professional associations, trade associations might very well want to communicate your concerns to the appropriate National District Attorneys Association, State Attorney Generals Association, National Association of Chiefs of Police, and I think you would find a very favorable and positive response.

Mr. COBLE. I thank the gentleman. I thank the panelists, I think those in the hearing room who are obviously interested in this issue. I apologize for our abrupt departure, but we must go vote. But again, I thank the witnesses for your testimony.

In order to ensure a full record and adequate consideration of this important issue, the record will be left open for additional submissions for 7 days. Also, any written question from a Member to the panel will be required within that same 7-day period.

This concludes the legislative hearing on H.R. 4239, the "Animal Enterprise Terrorist Act." We thank you for your cooperation and attendance. And without objection, the Subcommittee stands adjourned.

[Whereupon, at 11:25 a.m., the Subcommittee was adjourned.]

APPENDIX

Material Submitted for the Hearing Record

The Honorable Robert C. Scott, a Representative in Congress from the State of Virginia, and Ranking Member, Subcommittee on Crime, Terrorism, and Homeland Security

Statement
Congressman Robert C." Bobby" Scott
Ranking Member
Subcommittee on Crime, Terrorism and Homeland Security Hearing on
H.R. 4239, the "Animal Enterprise Terrorism Act"
April 23, 2006

Thank you, Mr. Chairman. And I would like to thank you for holding this markup on H.R. 4239, the "Animal Enterprise Terrorism Act". Apparently, current federal laws that were designed to protect businesses and employees in animal enterprises are doing a pretty good job. However, many of these businesses and employees are now complaining that other businesses and non-profits, and their employees, board members and family members, with whom they are affiliated are being stalked, harassed, intimidated, and have had their businesses, homes or cars vandalized, with some individuals even physically assaulted. The indications are that animal rights groups that use extreme tactics to press their point of view are taking advantage of the fact that the animal enterprise laws do not cover these type secondary relationships to wage a campaign of threats, harassment, intimidation and fear-mongering, in an effort to have them sever their relationships with targeted animal enterprises. This bill is designed to cover these perceived gaps or loopholes in the current animal enterprise protection laws.

Citizens engaging in lawful activities, as well as those associated with them, are entitled to be protected from criminal acts, and to be able to go about these lawful activities free from threats to their person or property, and that of their family and associates. State laws are generally good at providing those protections. However, the interstate nature of the planning and conduct of these type criminal and harassment tactics by groups skilled in avoiding the laws, make it difficult for states to effectively get at the some of the problems, and that's what this bill is designed to cover.

While we must protect those engaged in law animal enterprises, we must also protect the right of those engaged in their first amendment freedoms of expression regarding such enterprises. The issue was acknowledged and addressed in the bill. However, we have received concerns that the protections do not go far enough to assure First Amendment freedoms are not compromised. And I am aware, Mr. Chairman, of your proposals for further improvements in this area, and I want to work with you to ensure that we provide the protections of lawful activities needed here without jeopardizing First Amendment freedoms.

Included in those freedoms in my view, Mr. Chairman, is a right to engage in peaceful civil disobedience, and I am not sure the proposals adequately take that into account. If a group's intention were to stage a sit down, or lie down, to block traffic to a targeted facility, they certainly run the risk of arrest for whatever traffic, trespass or other laws they are breaking, but should not be held accountable for business losses due to delivery trucks being delayed, anymore than they should be for conducting a boycott of the business.

I am also concerned, Mr. Chairman, that the bill treats conspiracies and attempts the same as a completed offense. While someone who does not complete an offense solely because they were caught by law enforcement before completion should not be rewarded, I believe we should always encourage potential offenders to change their mind at anytime. Insisting that offenders who decide not to go through with a crime get the same sentence as if they had, only helps ring leaders or others promote the philosophy that "if I'm going to be shot for being a wolf, I might as well eat the sheep".

So, Mr. Chairman, I look forward to the testimony of our witnesses on how we strike a proper balance of protecting lawful activities and our First Amendment freedoms. Thank you.

Prepared Statement of the Honorable Sheila Jackson Lee, a Representative in Congress from the State of Texas

SHEILA JACKSON LEE
18TH DISTRICT, TEXAS

WASHINGTON OFFICE:
2435 Rayburn House Office Building
Washington, DC 20515
(202) 225-3816

DISTRICT OFFICE:
1919 SMITH STREET, SUITE 1180
THE GEORGE "MICKEY" LELAND FEDERAL BUILDING
HOUSTON, TX 77002
(713) 655-0050

ACRES HOME OFFICE:
6719 WEST MONTGOMERY, SUITE 204
HOUSTON, TX 77019
(713) 691-4882

HEIGHTS OFFICE:
420 WEST 19TH STREET
HOUSTON, TX 77008
(713) 961-4070

Congress of the United States
House of Representatives
Washington, DC 20515

COMMITTEES:
JUDICIARY
SUBCOMMITTEES:
CRIME, TERRORISM, AND HOMELAND SECURITY
RANKING MEMBER
IMMIGRATION, BORDER SECURITY, AND CLAIMS

HOMELAND SECURITY
SUBCOMMITTEES:
INTELLIGENCE, INFORMATION SHARING, AND TERRORISM RISK ASSESMENT
ECONOMIC SECURITY, INFRASTRUCTURE PROTECTION, AND CYBERSECURITY
MANAGEMENT, INTERGRATION, AND OVERSIGHT

SCIENCE
SUBCOMMITTEES:
ENERGY
SPACE AND AERONAUTICS

MEMBER
DEMOCRATIC CAUCUS POLICY AND STEERING COMMITTEE

CO-CHAIR
CONGRESSIONAL CHILDREN'S CAUCUS

Congresswoman Sheila Jackson Lee, of Texas

Statement on HR 4239, the Animal Enterprise Terrorism Act

May 23, 2006

Mr. Chairman, I appreciate the opportunity today to discuss the issues involved with HR 4239, the Animal Enterprise Terrorism Act.

The Animal Enterprise Terrorism Act rewrites federal criminal code provisions regarding animal enterprise terrorism to prohibit anyone from traveling in, or using the mail or any facility of, interstate or foreign commerce for the purpose of damaging or disrupting an animal enterprise and, in connection with such purpose: (1) intentionally damaging, disrupting, or causing the loss of property used by or owned in connection with such enterprise; (2) intentionally placing a person in reasonable fear of death or serious bodily injury to that person or a family member through threats, vandalism, property damage, trespass, harassment, or intimidation; or (3) conspiring or attempting to do so.

We are fortunate that we live in a country that protects one's rights to think and believe independently and freely. However, there is a very distinct line which some animal rights activist groups are alleged to have stepped across. It is important to recognize that another person's business, livelihood, property or freedom cannot be in danger due to the actions of a passionate individual.

However, in considering the varying perspectives today, I want to ensure that First Amendment rights are preserved. We should be careful not to criminalize behavior that is protected by our Bill of Rights.

At the conclusion of the hearing today, I hope to have an answer to one question that I consider critical: Is there a need for this, and why are existing laws not adequate?

Thank you, Mr. Chairman, and I yield the balance of my time.

Prepared Statement of the Honorable James Inhofe, U.S. Senator Committee on Environment and Public Works

Statement of the Honorable James Inhofe
Chairman, U.S. Senate Committee on Environment and Public Works

Before the Subcommittee on
Crime, Terrorism, and Homeland Security
Committee of the Judiciary
U.S. House of Representatives
Addressing Eco-terrorism
May 23, 2006

Today this Subcommittee will take up the issue of Eco-terrorism and what the Congress can do to address this dangerous movement in the United States. Eco-terrorism is not unlike any other form of terrorism, in which people with an agenda -- in this case, in the name of the environment or animals -- use force and violence to influence individuals, companies, and governments to adopt a particular policy. Animal rights extremists in particular use tactics such as bombings, arson, sabotage, stalking, and harassment, to frighten individuals and, in turn, the companies they work for into abandoning the use of animals for research. If this is not bad enough, the terror tactics are not limited to the companies that use animals for research but extend also to companies that do business with companies that use animals for research. This radical system of activism is called "tertiary targeting" and has unfortunately proven to be highly effective. Employees with nothing to do with research on animals, but work for a company that provides insurance or courier services or banking services to companies that conduct research on animals have been viciously targeted. In some cases, these employees have been watched in their homes and followed for weeks on end with their every move documented by the animal rights extremists and posted on their website. Information such as where their children go to school, what sports their children play and where, and the exact routes they drive to and from work are examples of the personal information that these activists post on their website with a call for their membership to "teach them a lesson."

Supplementing the intimidation and harassment tactics, groups like the Animal Liberation Front, the Earth Liberation Front ("ELF"), and Stop Huntingdon Animal Cruelty ("SHAC") have also been responsible for bombings in California and arsons across the country with over 1,200 acts of violence and $200 million in damages. As Chairman of the Committee on Environment and Public Works, I have held two hearings on Eco-terrorism and have received testimony from multiple victims, as well as the FBI, ATF, and DOJ, all of whom impressed upon the need for legislation to stop the horrible infringement of the rights of citizens who are just trying to make a lawful living. I even heard testimony from the animal rights groups. During my questioning of Dr. Jerry Vlasak, the spokesman for ALF, he actually defended a statement he made when he was speaking to an animal rights convention -- that the assassination of research scientists would be a good tactic to scare scientists away form conducting research on animals. Dr. Vlasak also believes that a mouse is "the moral equivalent to a child."

Belief systems like this -- that value a child equal to a mouse -- lead these extremists to relentlessly assault those they believe are mistreating animals. This includes scientists looking for cures to cancer, Alzheimer's disease, and hundreds of other conditions that take our loved ones from us. This is why I wrote and introduced Senate Bill S.1926, the Animal Enterprise Terrorism Act. This bill, which Congressman Petri introduced in the House -- H.R. 4239 -- will help protect those working to develop science by providing the necessary tools to federal law enforcement so that they may adequately investigate, apprehend, and prosecute these offenders. I urge this Committee to support this legislation and I thank Chairman Coble for your leadership on this issue in the House.

PREPARED STATEMENT OF THE HONORABLE THOMAS E. PETRI, A REPRESENTATIVE IN CONGRESS FROM THE STATE OF WISCONSIN

The Honorable Thomas E. Petri
Subcommittee on Crime, Terrorism, and Homeland Security
H.R. 4239 Animal Enterprise Terrorism Act of 2005
Statement for the Record, May 23, 2006

Chairman Coble, Ranking Member Scott and Members of the Subcommittee: I want to take this opportunity to thank you for holding this hearing on legislation I have introduced, the *Animal Enterprise Terrorism Act*. To date, H.R. 4239 has 32 cosponsors. I applaud your effort for taking on an issue that is so important to those in agricultural, biomedical, and biotechnology industries as well as to university research institutions.

Between January of 1990 and June of 2004, extremist movements such as the Animal Liberation Front (ALF), the Earth Liberation Front (ELF), and Stop Huntington Animal Cruelty (SHAC) committed more than 1,100 acts of terrorism causing more than $120 million in damages. Animal rights extremists advance their cause through "direct action", which includes death threats, vandalism, animal releases, and bombings. Their actions are calculated to aggressively intimidate and harass those identified as targets. Traditional targets include research and biomedical laboratories, fur farms, and restaurants. More recently, these extremists have also turned to targeting companies and individuals that do business or have a financial interest in an animal enterprise.

The FBI considers these extremist groups among its most serious domestic threats. Current federal law, including the Animal Enterprise Protection Act, is inadequate to address the threat posed by violent acts committed by these animal rights extremists. According to an official at the Justice Department, "... SHAC and other animal rights extremists have recognized limits and ambiguities in the statute and have tailored their campaign to exploit them."

In my own state of Wisconsin, mink farmers and biomedical researchers have experienced their own fair share of intimidation, harassment, and vandalism at the hands of animal rights extremists. Farmers have had their properties raided, causing thousands of dollars of damage. One of my first experiences with the destruction caused by animal rights extremists was the burning down of a feed mill in my district that caused $1.5 million dollars in damage. ALF claimed immediate responsibility for the incident and posted this message on its Web site: "You are not forgotten. More to Come." Many farmers continue to face the threat of violence via threatening letters, phone calls and emails.

As you will hear today, this violence and intimidation by animal rights extremists is not limited to the agricultural community in Wisconsin. Scientists around the state have received razor blades in the mail at their homes with letters stating they were laced with the AIDS virus. Personal information such as home addresses, phone numbers and photographs of researchers have been posted on extremists' Web sites. Many of these same scientists report death threats and home visits by animal rights extremists who through their terrorism have a goal of driving the scientists out of their research- research which has and will continue to improve human health and quality of life.

That is why I introduced the Animal Enterprise Terrorism Act of 2005 (H.R. 4239), which amends the Animal Enterprise Protection Act to provide federal authorities with the necessary tools to help prevent and better investigate and prosecute eco-terror cases. The legislation addresses gaps in the current law that keep authorities from using it in the most effective manner possible.

Drafted with technical assistance from the counter-terror experts at the Department of Justice and the Federal Bureau of Investigation, the bill broadens the definition of animal enterprise to include a commercial enterprise that uses or sells animals or animal products for profit including animal shelters, breeders, pet stores, and furriers. It provides for penalties for intentional economic disruption or damage and for intentionally causing bodily harm or placing a person in reasonable fear of death or bodily harm.

It also specifically addresses the "tertiary targeting" tactic employed by these extremists by prohibiting intimidation and harassment as well as intentional damage of property belonging to a person or organization with ties to an animal enterprise. Enactment of this legislation will enhance the ability of law enforcement and the Justice Department to protect law-abiding American citizens from violence and the threat of violence posed by animal rights extremists.

Some have raised questions as to the impact on First Amendment rights. I do not believe that the legislation infringes on any rights guaranteed under the Constitution. However, if the Subcommittee feels that changes need to be made to ensure that legal expression is not criminalized than I fully support these efforts.

Again, I want to thank the Subcommittee for holding this hearing. I hope that the testimony today will convince you that those who work in animal enterprises or have a connection to animal enterprises are under attack and need our protection.

Prepared Statement of Mr. Frankie L. Trull, President, National Association for Biomedical Research

H.R. 4239
THE ANIMAL ENTERPRISE TERRORISM ACT

STATEMENT OF FRANKIE L. TRULL

President, National Association for Biomedical Research

BEFORE THE HOUSE COMMITTEE ON THE JUDICIARY
SUBCOMMITTEE ON CRIME, TERRORISM, and HOMELAND SECURITY

May 23, 2006

Mr. Chairman and Members of the Committee,

On behalf of the membership of the National Association for Biomedical Research (NABR), I thank you for conducting this hearing on H.R. 4239, the Animal Enterprise Protection Act. Animal and eco-terrorism is a growing problem and is becoming increasingly aggressive and hostile. Your leadership is both desperately needed and greatly appreciated. I also want to acknowledge today's witnesses for the courage they have exhibited as they are putting themselves at considerable risk by speaking out on this issue.

The National Association for Biomedical Research (NABR) is the only national, nonprofit organization dedicated solely to advocating sound public policy that recognizes the vital role of humane animal use in biomedical research, higher education and product safety testing. Founded in 1979, NABR provides the unified voice for the scientific community on legislative and regulatory matters affecting laboratory animal research. NABR's membership is comprised of hundreds of public and private universities, medical and veterinary schools, teaching hospitals, patient groups, professional societies, pharmaceutical companies, and other animal research-related firms.

Animal research has played a vital role in virtually every major medical advance of the last century – for both human and animal health. From antibiotics to blood transfusions, from dialysis to organ transplantation, from vaccinations to chemotherapy, bypass surgery and joint replacement, practically every present-day protocol for the prevention, treatment, cure and control of disease, pain and suffering is based on knowledge attained through research with animals. Ample proof of the success of animal research can be found in the vast body of Nobel Prize winning work in physiology and medicine. Seven out of the last 10 Nobel Prizes in medicine and 69 awarded since 1901 have relied, at least in part, on animal research

In fact, research on animals is in many cases an obligation. According to the Nuremburg Code, drawn up after World War II as a result of Nazi atrocities, any research on humans "should be designed and based on the results of animal experimentation." The

Declaration of Helsinki, adopted in 1964 by the 18th World Medical Assembly and revised in 1975, also states that medical research on human subjects "should be based on adequately performed laboratory and animal experimentation." As well, the FDA expressly requires that laboratory animal tests be conducted both for prescription drugs and over-the-counter drugs before these products can be tested further in humans.

Since its inception, NABR has witnessed many changes in animal rights activism. What began as a grassroots movement has grown into a sophisticated and often violent industry. The increased willingness of some animal rights extremists to use violence and to inflict economic and physical damage on any person or entity remotely associated with an organization that uses animals in research, has become a serious threat to the biomedical enterprise.

Violent acts committed in the name of animal rights have been carried out in this country for more than two decades. In the past, targets have consisted primarily of research facilities and companies, as well as researchers and their families. Congress responded to animal rights violence by enacting the Animal Enterprise Protection Act of 1992, codified as 18 USC 43 of the Criminal Code. This act made it a federal crime to intentionally cause physical disruption to an animal enterprise by stealing, damaging, or causing the loss of property used by an animal enterprise if these acts resulted in damages exceeding $10,000. The Act was amended in 1996 and again in 2002. The 2002 amendments made several improvements to the original Act, including making it a federal crime to engage in criminal acts resulting in damage of less than $10,000.

THIRD PARTY or TERTIARY TARGETING

Unfortunately, the Animal Enterprise Protection Act continues to be of limited use to federal law enforcement officials in combating violent and disruptive acts of animal rights extremism. Violent activists have employed a disturbing new strategy. Tactics still include arson, death threats, sabotage and vandalism, but the new approach is something called "tertiary" or third-party targeting. It is this type of targeting that the original Animal Enterprise Protection Act and its subsequent amendments did not envision. Consequently, law enforcement has very limited means to protect third parties from the actions of animal rights extremists. By aggressively targeting clients, insurance companies, banks, health providers, accounting firms, shareholders, market makers, internet providers, even lawn care, catering and other service companies, activists have found an effective way to disrupt the financial health and functioning of companies engaged in animal research.

The most successful proponent of tertiary targeting has been a UK-born group called Stop Huntingdon Animal Cruelty (SHAC). SHAC has successfully targeted third-parties since the late 1990's in its campaign against Huntingdon Life Sciences (HLS), a large contract research firm. Its targets have included some of the best known financial services companies in the world and the campaign has succeeded to the extent that the UK government has been forced to act as the banker and insurer for Huntingdon Life

Sciences. U.S. animal extremists have learned well from their UK colleagues and many of the tactics perfected overseas have now been employed against American targets.

Just to be clear, I am not referring to tactics aimed at tertiary targets that involve the use of picketing, boycotts, letters, phone calls, letters to the editor, advocacy of new laws and regulations, or other forms of legal protest that are protected by the First Amendment. It is the threat of physical violence, property damage, intimidation, coercion, and harassment that are the key weapons of these campaigns.

A couple of examples may help illustrate animal extremist tactics:

- electronic denial of service attacks where a handful of activists using a computer program anywhere in the world can bombard a web site or email system with so much information that it crashes;
- phone auto-dialers where activists using a computer call company numbers hundreds of times a day, effectively tying up a company's phone system;
- black faxes, where endless sheets of black paper are sent to a fax machine causing it to burn out;
- letters to companies threatening consequences, and citing examples, if they do not cease doing business with animal research companies;
- theft of personal information like home phone numbers, credit card numbers, bank account numbers, and social security numbers of company employees and their neighbors, where the information is then posted on the Internet;
- "home visits" where activists visit homes in the middle of the night with bullhorns and distribute "wanted for murder" posters to neighbors;
- smoke bombs set off in office towers, causing the evacuation of hundreds of employees;
- death threats against researchers, company employees, and their families;
- vandalism and destruction of property such as cars, bank machines, locks and windows;
- office invasions, where activists protest outside an office, and then rush in to occupy the facility to steal documents, destroy offices and assault employees.

On August 28, 2003, the campaign against HLS produced a frightening new twist: bombings. Two pipe bombs were set off outside of Chiron Corporation in Emeryville, California. The first went off in the early morning hours, but the second was deliberately set for half an hour after the first, designed to harm the first responders. Chiron had at one time been a client of HLS and was listed as a target on SHAC's Web site.

On September 26, 2003 a second set of pipe bombs, wrapped in nails, were set off at the Shaklee Corp. facility in Pleasanton, California. Shaklee is a subsidiary of a Japanese company that activists have tied to HLS. It is by sheer luck that there were no injuries in either of the blasts.

Responsibility for the bombings was claimed by a previously unknown group calling itself "The Revolutionary Cells for Animal Liberation." But there appears to be an interrelation between activists willing to carry out acts of violence and long, unrelenting campaigns of intimidation and harassment. SHAC, which according to the FBI has an "extensive history of violence" uses its Web site to post lists of targets, including the bombing targets Chiron and Shaklee. Those target lists include the home phone numbers and addresses of executives and employees of targeted companies. Groups advocating "direct action" like SHAC and the Animal Liberation Front (ALF) also seem to have leaders in common. For example, Kevin Kjonaas (or Jonas) who speaks for SHAC USA, and was recently convicted on federal charges related to animal extremism, was at one time spokesperson for the ALF.

As a result of these campaigns, not only are the rights of companies to freely do business being compromised, but security costs are soaring both for private companies and public colleges and universities. Money that could be directed at researching cures and treatments for disease is being re-directed to provide extra security for existing research. Many companies have been forced to hire personal security to protect the homes of their employees.

More often than not, these extremists claim that they are exercising their right to free speech. I want to make it very clear that NABR and its members fully support constitutionally-protected rights to free speech. However, coordinated campaigns that include threats, intimidation, coercion, harassment, and other tactics that place people in fear of physical harm to themselves or their friends and families are not forms of protected free speech. These are the tactics that extremist groups are using to forcibly impose their will on our law-abiding organizations, and we urge the Congress to take action by providing federal law enforcement with adequate tools to prosecute those who violate the rights of others.

For many years, our members have sought ways to protect their institutions against the threat of animal rights terrorism. NABR has long been active in working with Congress to find ways of effectively addressing this increasingly complex problem. Now, we find that current laws are inadequate to address the new tactics being employed by animal extremists. In fact, these campaigns seem to be designed to skirt existing laws.

We urge the Committee to act on H.R. 4239, the Animal Enterprise Terrorism Act to protect our members from the evolving tactics of animal rights extremists. The continuation of life-saving medical research, the safety of researchers and their families, and the economic health of this important industry, depends on our finding effective and immediate ways to address this problem. Law enforcement needs new tools to pursue and prosecute those who are perpetrating these violent, organized, and methodical campaigns against institutions that conduct animal research and third parties that do business with them. Our members are urging us to deliver this message to Congress – to eliminate this climate of fear.

Thank you again, Mr. Chairman, for allowing me to submit testimony for the record, and for holding this important hearing.

Prepared Statement of Dr. Bruce R. Bistrian, President of the Federation of American Societies for Experimental Biology (FASEB)

Testimony of the Federation of American Societies for Experimental Biology (FASEB)
President Bruce R. Bistrian, M.D., Ph.D.
House Committee on the Judiciary
Subcommittee on Crime, Terrorism & Homeland Security
May 23, 2006

The Federation of American Societies for Experimental Biology (FASEB) appreciates the opportunity to submit testimony on behalf of our 22 scientific society members, representing more than 84,000 biomedical research scientists. Furthermore, FASEB gratefully acknowledges the leadership role of subcommittee Chairman Howard Coble and full committee Chairman James Sensenbrenner for holding this hearing to consider an important piece of legislation and to highlight the detrimental consequences of animal rights (AR) extremism. FASEB believes it is important to provide the statutory tools necessary to protect those who have been targeted as victims by groups who advocate arson, personal violence, and vandalism as appropriate means to an end. Therefore, we stand in support of H.R. 4239, the Animal Enterprise Terrorism Act of 2005.

The use of animal models in biomedical research is absolutely essential to our ability to develop treatments and cures for those suffering from debilitating diseases. Breakthroughs in treatments for diabetes, heart disease, cancer, HIV / AIDS and Parkinson's disease would not have been possible without the use of animals. Moreover, animal research directly benefits animals themselves: the majority of veterinary advances are a direct result of research involving animal models.

Members of the FASEB Societies believe that the use of animals in research and education is a privilege. This imposes a major responsibility to provide for their proper care, ethical and humane treatment. Good animal care and good science go hand-in-hand and is taken most seriously by the scientific community. In addition, FASEB feels that it is a responsibility of researchers to communicate to the public about the role and importance of animals in research, a task made difficult due to the dangers posed by members of AR extremist groups.

The recent escalation in violence and intimidation campaigns by AR extremists directed towards researchers and their institutions is of great concern to our members. Biomedical researchers are often the focus of campaigns launched by groups like Stop Huntington Animal Cruelty (SHAC) or the Animal Liberation Front (ALF). We would like to take this opportunity to put a human face on researchers whose lives have come under siege. These are scientists who have dedicated decades of their life in an effort to alleviate human suffering and improve human health. Moreover, these incidences are often widely reported when they take place and are counted as victories by the AR extremist movement.

In the November 22, 2002 edition of *Science* magazine[1], then FASEB President Steven Teitelbaum, MD, published an Op-Ed in response to the ongoing situation of Michael Podell, DVM. Dr. Michael Podell was an Associate Professor in the Department of Veterinary Clinical Sciences and Center for Retrovirus Research, College of Veterinary Medicine at the Ohio State University (OSU). He studied the effect of methamphetamine abuse on the progression of immunodeficiency virus, known in humans as HIV, the causative agent of AIDS, as well as the neurological effects of HIV, a poorly understood aspect of the disease. Podell's investigation involved the use of feline models, because his previous research had discovered that feline

[1] Teitelbaum, S. (2002, Nov. 22) *Science*. 298:1515.

immunodeficiency virus (FIV) closely mirrors the neurodegenerative effects of HIV infection in humans, making cats an excellent surrogate for HIV neuropathology. This research was uncovering tantalizing new evidence about the effects of methamphetamine use on viral replication. Podell's findings, published in the *Journal of NeuroVirology*, plainly showed a manifold increase in neural cells' ability to replicate FIV after methamphetamine treatment. The OSU study also shed light on the mechanism by which FIV associates with astrocytes, mutating into a strain of virus that does not depend on immune system interaction for replication. This important piece in the puzzle of how the viral load of FIV/HIV in the brain leads to dementia is vital to efforts to lessen this debilitating brain damage. However, before these findings were published, Dr. Podell abandoned his research, walking away from a $1.68 million dollar grant from the National Institutes of Health (NIH), the sort of funding only one in five researchers might successfully apply for.

Why did Michael Podell, a promising and successful scientist who was contributing invaluable knowledge to the fight against AIDS, leave his career as a research scientist? From the awarding of the grant in October, 2000, Dr. Podell, his family, and Ohio State University were subjected to an intense and often violent campaign of harassment. According to interviews with Dr. Podell[2], he received thousands of harassing emails and a dozen death threats. The university itself was subjected to repeated acts of vandalism. Ultimately, it was concern for the safety of his family, who were also being threatened, that played a large part in convincing Michael Podell to leave. In one threatening letter they received, a newspaper picture of a United Kingdom

[2] Davis, S. (2002, August 1) *DVM Newsmagazine*.

scientist whose car had been bombed by AR extremists was scrawled with the message "You're next.[3]"

Sadly, this is not an uncommon story: throughout the US and UK, law-abiding biomedical researchers are being targeted. Although these groups have sought cover under the fact that no US researcher has been physically harmed or directly targeted for physical violence, this is not true in the UK. Human targets of groups like SHAC and ALF in Europe have been beaten, branded, attacked with caustic substances, and firebombed[4]. The roots of the US movement of AR extremism are in the UK. FASEB feels it is only a matter of time before these domestic campaigns escalate to the violent intensity of their UK counterparts. We fear that recent events, such as the New York Stock Exchange's apparent capitulation to threats by AR extremists in canceling the stock listing of Huntington Life Sciences, will embolden these groups to increase their violent efforts to halt the use of animals in research.

Research institutions, funded in large part by taxpayer dollars, are also victims of AR extremism. In the last hearing on this topic held by this committee, you examined the damage done during the November, 2004 attack at the University of Iowa, which caused $450,000 worth of damage to laboratories and equipment. Again, this is not an uncommon story: Louisiana State University (LSU) in Baton Rouge was spared the wrath of Hurricane Katrina, but has been attacked multiple times by ALF[5], most recently in April, 2005[6]. On ALF's website, where the attack against LSU is triumphantly detailed,

[3] Stolberg, SG (2002, July 23) *New York Times*. P. A1
[4] Bhattacharya, S. (2004, April 22) *The New Scientist*. Retrieved 2005, Oct. 25 from: http://www.newscientist.com/article.ns?id=dn4913
[5] McElfresh, A. (2004, Jan. 30) *The Reveille (LSU)*. Retrieved 2005, Oct. 25 from: http://www.lsureveille.com/vnews/display.v/ART/2004/01/30/4019fdcbb1286
[6] Smallwood, S. (2005, Aug. 5) *The Chronicle of Higher Education*. 51(48). P. A8.

there is an ominous message to researchers, "Stop now, or be stopped.[7]" Money that could be going towards life-saving medical research is now being spent on increased security and cleaning up the damage caused by AR extremists.

But the monetary damages done to laboratories and research institutions, (documented by the Federal Bureau of Investigation, the Southern Poverty Law Center, the National Association of State Universities and Land-Grant Colleges, and the Foundation for Biomedical Research), do not fully convey the impact of direct actions by ALF and SHAC. The loss of computer files, lab animals, research notebooks or microscope slides may not account for a great monetary loss, but could represent years of work in the life of a scientist or graduate student. Imagine working long hours on a small stipend, following your heart's passion through the sometimes frustrating process of bench research for five years, trying to achieve a doctoral degree, only to have all of that work eliminated in one night by a group whose public website describes in great detail exactly how to attack and harm research laboratories. The human toll of having one's research lab targeted by AR extremists is described eloquently in a *Washington Post* editorial (July 17, 2005) written by Dr. Mark Blumberg, a researcher at the University of Iowa, which we have attached.

Animal rights extremists have become a serious impediment to the progress of biomedical research, as well as to the peace of mind of scientists themselves. We urge the committee to carefully examine the toll of AR extremism on the scientific community and our medical research enterprise, the results of which are an insidious stifling effect on the discovery of new cures for diseases. Biomedical researchers are dedicated to

[7] Animal Liberation Front. Retrieved 2005, Oct. 25 from: http://www.animalliberationfront.com/ALFront/Actions-USA/LSU_Rat_Lib_05.htm

improving the health, well-being and productivity of all people. They should be allowed to do so in safety and security. Thank you for the opportunity to submit testimony and for shedding light on this critical issue. FASEB stands ready to assist in any way possible.

washingtonpost.com

The Animal Zealotry That Destroyed Our Lab

By Mark S. Blumberg
Post
Sunday, July 17, 2005; B03

http://www.washingtonpost.com/wp-dyn/content/article/2005/07/15/AR2005071502181_pf.html

IOWA CITY, Iowa

"Are you lying down?" my wife asked me over the phone. It was Sunday, Nov. 14 of last year, and I was just waking up in my hotel room in Madison, Wis., where I'd gone to visit my sister and her son for the weekend. My wife's question -- especiallyher urgent tone-- triggered a cascade of sickening thoughts. Soon, I was racing home to Iowa.

Although the pieces only came together over the next several days, the bare facts were these: Early that morning, at least five individuals had illegally entered the research facility at the University of Iowa where my colleagues and I, all professors of psychology and neuroscience, work. The intruders broke into offices and laboratories, dumped acid and other chemicals and destroyed equipment. They also "liberated" the animals -- primarily rats and mice -- used in our studies of such basic behavioral and biological processes as learning, memory, temperature regulation and sleep. One of my graduate students arrived at work early that morning and discovered, in bold red spray paint, the slogans that are the hallmark of the Animal Liberation Front (ALF): "Science not sadism" and "Free the animals."

With this break-in, my department had become thelatest poster child of the animal rights movement. After years of escalating attacks on research facilities in the United Kingdom, animal rights and environmental extremists have turned to North America, which is fast becoming a breeding ground for their type of violence. But because the number of individuals affected is still relatively small, most Americans remain unaware of the seriousness of the threats. As my experience shows, even among decision-makers, few are taking it seriously enough.

The care of laboratory animals isn't, as some seem to believe, an unregulated field. As scientists engaged in government-sponsored research, we must conform to an exhaustive array of local, state and federal rules. Nor are we unthinking about these animals' use. As scientists, we debate it among ourselves and with others, as all thoughtful individuals do when dealing with issues of life and death. What happened in Iowa, though, was not a debate; it was an assault.

For us, the break-in set off a chain of events that one might expect after an attack of such magnitude. Our unassuming buildings at the edge of campus were cordoned off as local, state and then federal law enforcement personnel descended. With the closing of these buildings, the daily lives of hundreds of faculty, staff and students were disrupted. Experts in the handling of hazardous materials spent weeks identifying and removing the corrosive chemicals that had been dumped inside.

The cost of the cleanup, replacement of valuable equipment and purchasing of new animals totaled in the hundreds of thousands of dollars. Contrary to initial reports, relatively little data were lost (in part because the attackers seemed more concerned with smashing computers than erasing hard drives) although even small losses can have far-reaching consequences for research.

Instead, it was the human cost that was most devastating. Imagine the horror of walking into your office at work, as one of my young colleagues did, to find computers, books and personal effects (such as ultrasound images of your unborn child) soaked in acid. Then, imagine having to don a chemical protection suit for several days and sift through multiple 55-gallon drums filled with acid-soaked papers, photocopying those that are still readable as they crumble in your hand.

Unfortunately, the attack on the building is where our story begins, not ends. For what followed was a series of well-orchestrated harassments. First came the e-mailing of a communique to the media, detailing the crime and the rationale for targeting our facility and the individuals who work there. Each of us was singled out for derision; I was colorfully described as having a "famously deranged mind" because of my research on the similarities between the high-pitched squeals of infant rats and the life-sustaining grunts of human preemies in respiratory distress.

Some of ALF's statements produced the desired chilling effect: "Let this message be clear to all who victimize the innocent," the e-mail read. "We're watching. And by axe, drill, or crowbar -- we're coming though your door. Stop or be stopped." Later in that document, the brazen and indiscriminate nature of their threat was revealed when, after noting "the established link between violence towards animals and that towards humans," they listed "as a public safety measure" our names, our spouse's names, home addresses and phone numbers, as well as information about our students.

Next came the video. Several days after the communique, local journalists informed a group of us that a surreptitious delivery had brought a 50-minute videotape of the crime. Would we be interested in seeing it? Within an hour, two colleagues and I found ourselves huddled together in front of a small television set in a local newsroom, watching in dismay as these individuals -- clearly youthful despite being hidden behind hoods, masks and gloves -- paraded through our facility, smashing delicate instruments with oversize hammers and transferring rats and mice to plastic cages. It was particularly difficult for me to watch as my infant rats, along with their mothers, were thrown together with several other adults, knowing (as these animal "liberators" apparently did not) that cannibalism of the young was the likely outcome. There was no video of that.

In the weeks thereafter, our attackers and their allies kept up their campaign. There were press conferences by local agitators, freedom of information requests, midnight phone calls, a well-publicized visit by a nationally known pro-ALF speaker whose message was that more attacks were needed. And then came the magazines. They started as a trickle, but soon my mailbox was deluged with dozens catering to every taste: Canoe & Kayak, Guns & Ammo, Fit Pregnancy, Muscle Mustangs & Fast Fords. It's simple but ingenious: tear out those little subscription cards, apply a label, and send it in. No hassle, no mess. In total, nearly 450 subscriptions were directed at us, 160 to me alone. Funny? Perhaps, unless you consider how you would respond to such an onslaught, including the invoices and, ultimately, the credit agencies that followed.

When we learned that a Senate panel would be addressing the issue of animal rights extremism in May, we thought that some relief was imminent. Groups like the Southern Poverty Law Center and Anti-Defamation League have been keeping an eye on the growing violence. Critics have pointed out financial donations, overlapping personnel and supportive public statements that raise questions about a possible relationship between above-ground groups such as People for the Ethical Treatment of Animals (PETA) and fringe groups like ALF. We hoped that such evidence had accumulated to the point that a concerted and bipartisan effort might finally affect their formidable fundraising apparatus. We were sadly disappointed.

We were encouraged that the president of our university had been called as a witness and that our experiences of the past several months would receive some high-level attention. Unfortunately, the hearing quickly devolved into a partisan disagreement. Incredibly, the senators seemed more interested in protecting their favored activist groups from scrutiny than in determining which groups actually posed significant threats to the lives and livelihoods of law-abiding citizens. Most galling were the comments of Sen. Frank Lautenberg, a Democrat from New Jersey, who seemed miffed that his time was being wasted on such fluff. Incredulous of the testimony provided by the FBI and the Bureau of Alcohol, Tobacco, Firearms and Explosives (ATF), in which violent animal rights and environmental extremists were identified as among our most serious domestic terrorism threats, Lautenberg asked facetiously who the next target would be: "Right to Life? Sierra Club?" Then, he inexplicably proclaimed himself "a tree hugger."

I later made several attempts to contact Lautenberg about his comments, via fax, phone and e-mail, but never received a response.

I was a victim of a violent crime once before. While on break from college in the early 1980s, I was sitting in my parents' home in Chevy Chase reading a book when, suddenly, I looked up and found myself staring into the barrels of two snub-nosed revolvers. The intruders tied me up and robbed the house, then left silently. As traumatic as that event was, its effect on me was fleeting. I was angry, yes, but I did not feel terrorized. These home invaders clearly did not hate me for who I was or what I did. They did not issue a communique declaring that others should attack me. They did not release a video to force me to relive the indignity of the event. And they did not encourage their minions to engage in further harassments. Terrorists, no matter what their cause, seek political change through violence and intimidation. Is it essential that we label animal rights extremists as terrorists? Perhaps not, unless such a label helps us -- and especially politicians -- to better appreciate the seriousness of the threat and to marshal the necessary law enforcement resources.

Because the threat is serious. Today, scientists, clinicians and educators find themselves engaged in a seemingly-endless string of pitched battles: over the teaching of intelligent design in our public school classrooms, over the availability of stem cells to treat degenerative diseases, over the rights of severely brain-damaged individuals to die. If we focus on the conventional politics that drive these conflicts -- right vs. left -- we miss the bigger picture.

In fact, what ties all of them together is a common distrust of and disdain for science, for empirically based medicine, for the value of evidence and critical analysis, and for progress in a free and open society. Moreover, and perhaps most alarming, is the adoption by certain groups of increasingly violent action to achieve their political aims. Indeed, the mounting acceptance of intimidation and violence within the anti-abortion movement eerily parallels the escalating tactics of animal rights extremists. Thus, the ideology and goals of these groups may align at opposite ends of the political spectrum, but their tactics have converged. As we know, a number of abortion doctors have already been killed, and some animal rights extremists seem to approve of physical violence as a tactic . It's only a matter of time before someone takes the next step. Whom will Lautenberg hug then?

Author's e-mail: msblum24@hotmail.com

Mark Blumberg is a behavioral neuroscientist at the University of Iowa. He is the author of "Body Heat: Temperature and Life on Earth" (Harvard University Press) and the forthcoming "Basic Instinct: The Genesis of Behavior" (Thunder's Mouth Press).

PREPARED STATEMENT OF MR. MARK L. BIBI, GENERAL COUNSEL, LIFE SCIENCES RESEARCH

WRITTEN TESTIMONY OF MARK L. BIBI, GENERAL COUNSEL OF LIFE SCIENCES RESEARCH, INC. AND HUNTINGDON LIFE SCIENCES INC., FOR THE HOUSE JUDICIARY COMMITTEE SUBCOMMITTEE ON CRIME, TERRORISM AND HOMELAND SECURITY

HEARINGS ON H.R. 4239, THE "ANIMAL ENTERPRISE TERRORISM ACT"

MAY 23, 2006

Chairman Coble, Ranking Member Scott and members of the Subcommittee:

My name is Mark Bibi. I am General Counsel of Life Sciences Research and its operating subsidiary, Huntingdon Life Sciences. Thank you for this opportunity to submit testimony regarding H.R. 4239, the Animal Enterprise Terrorism Act. Passage of that bill is urgently needed by the medical research, pharmaceutical, agricultural and biotechnology industries, as well as by our research universities who conduct vital, life saving animal based research. This bill is needed to stem the growing dangers posed by animal extremist groups like Stop Huntingdon Animal Cruelty (known as SHAC), Win Animal Rights (known as WAR), the Animal Defense League, the Animal Liberation Front and their supporters. These groups have been identified by the FBI as one of the nation's leading domestic terrorist threats. It is a serious matter and deserves a serious response. HR 4239 is the right response at the right time.

LSR is the publicly traded parent company of HLS, headquartered near Princeton, New Jersey. We conduct government-required animal testing on drugs and chemicals to identify risks to humans, animals and the environment. Because of these efforts to make sure products are safe, HLS and those who do business with us have been relentlessly terrorized by a brazen collection of self appointed animal saviors, a group which values animal life over human life.

HLS has been the main target of these groups for seven years. Using "any means necessary" (their words, not mine), including violence, intimidation, threats, stalking and harassment, their goal has been to drive HLS out of business. This is only a first step to their ultimate goal of full "animal liberation", which would mean the closure of all animal testing labs, all farms, all zoos and in fact ANY entity that uses animals. At first, the main victims were people like me – the employees and scientists who work at HLS. But the animal extremists eventually realized that HLS employees would stand up to their intimidation and would not yield, because we so strongly believe in what we do – we help bring life saving medicines to market. It was at that point that the animal extremists began utilizing a new and enormously successful strategy of targeting third parties who do business with HLS –and even more tangentially, third parties who do business with those third parties.

What these extremist groups have discovered is that third parties who provide services to HLS are more likely to yield to the intimidation, and sever their business relationship, than they are to put up with the continued harassment. Time and again, in dozens of cases, HLS' providers –from accounting firms, to banks, to lawn gardeners and even our security firm –have capitulated to SHAC's demand that they cease working with HLS, deciding it's easier to give in than to suffer the continued and escalating harassment and intimidation. SHAC and its ilk have succeeded in driving a veritable who's who of the US and worldwide business community into abandoning HLS – companies like Citigroup, Merrill Lynch, Deloitte & Touche, Royal Bank of Scotland, Goldman Sachs, Marsh & McLennan, Stephens Group, Charles Schwab and Bank of America. More than 40 market makers – companies that trade in LSR stock – have been targeted and driven away. Virtually all of LSR's institutional investors have been harassed into selling their LSR stock. These companies are victims, just as HLS is a victim.

The difference, as things stand today, is that HLS is protected under current law, but these third parties are not. The existing statute, the Animal Enterprise Protection Act, which was enacted in 1992, protects entities defined as "animal enterprises", which include animal research labs like HLS. However, the third parties doing business with that animal enterprise are NOT directly protected under current law. HR 4239 would fix that, and would protect those third parties from the reign of terror they are currently experiencing.

While I of course am disappointed that so many companies have yielded to the intimidation of the animal terrorists and have stopped doing business with HLS, I can understand how truly frightening it is to be targeted and victimized in this fashion, having lived with it for years. My home has been vandalized; my car has had a rock thrown through its windshield. "Puppy killer" has been spray painted on my house. I have received threats in the mail, on the phone and through e-mail. Others at HLS have suffered beatings, acid attacks, car and letter bombings. The impact of this violence -- and the implicit threat of future violence -- is a terrifying life-changing event.

The tactics being utilized by these extremist groups are increasingly despicable. The 90 year old mother of an LSR market maker was targeted at her assisted living facility in the Midwest. She was falsely told her son is a pedophile, and a hearse was sent to collect her body from the facility -- while she was still alive. In England, the grave of the mother-in-law of an animal breeder was desecrated and the body was stolen. The 7 year old son of a Marsh executive that had been harassed and intimidated for weeks by SHAC was at home with his mother when the door bell rang. The young boy ran to the kitchen, grabbed a kitchen knife and crouched behind the front door, whispering to his mother – "don't worry Mommy, I'll protect you from the animal people."

All of this, and more, has been done to third parties. They need help in order to be able to stand up to these tactics. I enjoy the protections of the law, since I work for an animal enterprise covered by existing law; they still don't. That is why passing HR 4239 is so crucially important.

The risks posed by SHAC and its ilk should not be underestimated. They are the tip of the iceberg, the test case for a whole new brand of activism through intimidation. Other activist campaigns are no doubt waiting in the wings to see how this growing threat is dealt with. Imagine the impact if SHAC tactics were used by those opposed to various other industries from defense, to mining, to oil, to timber, to who knows what else.

England is widely recognized as the birthplace of the animal extremist movement and many of the US animal rights extremists have trained in the UK. England has acknowledged the seriousness of the threat posed by these criminals, and last summer passed legislation remarkably similar in its scope and purpose to HR 4239. It has enabled the police in England to successfully arrest a number of animal rights extremists over the past year for the types of actions that previously would have gone unchallenged. Prime Minister Blair wrote an article in the London Sunday Telegraph on May 14 in which he stressed the need for civilized society to stand up to the animal terrorists. He remarked on the success of last summer's law and wrote "If more measures are needed to protect individuals, universities and firms to root out the criminal extremist fringe, we will provide them."

Honorable members of the Subcommittee, more laws ARE needed here in the US to root out this criminal extremist fringe. We cannot allow the domestic terrorism practiced, fostered and encouraged by SHAC and its followers to flourish in our own backyard. I urge you to speedily pass HR 4239.

Thank you.

Prepared Statement of Ms. Gale Davy, Executive Director, Wisconsin Association for Biomedical Research & Education

HR 4239, *The Animal Enterprise Terrorism Act*
Testimony
Submitted to the Subcommittee on Crime, Terrorism, and Homeland Security
May 23, 2006
By
The Wisconsin Association for Biomedical Research & Education

We would like to take this opportunity to thank Chairman Coble and the other members of the Subcommittee for addressing the critical issue of terrorism against animal enterprise.

The Wisconsin Association for Biomedical Research & Education is a non-profit public education organization representing bioscience research institutions, companies, healthcare organizations and science professionals in the State of Wisconsin. We urge you to support the *Animal Enterprise Terrorism Act*, sending a strong message that illegal acts designed to terrorize scientists and others directly or indirectly involved in animal research, education and testing will not be tolerated by our society.

More importantly, we urge you to support this bill as it would expand the definition of animal enterprise currently in place under the *Animal Enterprise Protection Act*, afford protections to lawful businesses and organizations not directly involved in animal research that become targets of animal rights terrorism, and provide law enforcement with the tools necessary to gather evidence on illegal activities.

Scientists and companies involved in critical biomedical research, education and testing deserve the same protections and standards of enforcement as other enterprises. Lawful businesses and organizations which support and do business with entities involved in animal research, education and testing also deserve these protections. The same tools available to protect the automotive industry, allowing federal law enforcement to track illegal chop shop operations, should be available to protect scientific enterprise. Strengthening the ability of federal law enforcement agencies to track illegal activities across state lines will also lessen the burden on local law enforcement.

Scientists and other professionals involved in research, education and testing in Wisconsin have suffered terrorism, harassment and intimidation by animal rights extremists. Despite that, the biomedical research community is protective of first amendment rights, as are most Wisconsinites. We do not seek an erosion of first amendment guarantees. Rather we are asking, through this bill, that scientists and others be better protected - through enhanced enforcement and investigative powers - from illegal activities. The true threat to our constitutional guarantees are the kinds of terrorism, threats and other forms of intimidation employed by animal rights extremists against lawful and ethical use of animals by scientists and other professionals.

In preparing this testimony, we asked Wisconsin scientists and other professionals involved in animal research, education and testing in our state, to share their personal stories about the attacks they have suffered. More than twenty Wisconsinites shared their stories with us. Most of them asked that we do not reveal their names in this testimony lest they become more acute targets.

It is difficult to argue against their wish for anonymity. The fear is real and justified. In previous Senate testimony on this issue, a former spokesperson for the Animal Liberation Front - a terrorist organization - stated that murder was a "morally justifiable solution to the problem." Several animal rights extremist websites have called for the "death" and "elimination" of scientists using animals. Some of our Wisconsin scientists have been named on these websites.

In light of this, we have elected to protect the anonymity of all of the victims who have shared their information with us in this testimony. Some scientists have chosen and will choose to speak publicly about their experiences and we applaud them. Should any member of Congress or their staff wish to meet with the scientists discussed and quoted in this testimony, we will gladly set up those private meetings.

During the past several years Wisconsin scientists have been terrorized, threatened and harassed by animal rights extremists. Unfortunately, Wisconsin is not special in this respect. Scientists nationwide and worldwide have suffered the same and worse at the hands of extremists.

Nine Wisconsin scientists received letters with razor blades, designed to inflict harm upon opening. In several instances these letters stated that the blade was poisoned with the AIDS virus. One scientist recounted his agony over having to explain to his children why they were not allowed to touch the mail. Another described how the letter affected her daily life, now full of worry for her young sons and her husband.

At least twenty Wisconsin scientists have received death threats. The wife of one prominent scientist received death threats by mail and telephone. While the scientist was away attending scientific meetings, the phone would ring constantly throughout the night. When his wife answered the phone, loud, electronic sounds would play or individuals would make threats against the scientist and against his wife.

Educators who use animals in medical education and training have been subjected to threats and barrages of email - as many as 500 in a day. Three individual educators involved with education laboratories have received death threat. One of these had his name, wife's name, home address and phone number listed on a website that called for his "extermination."

Posting of names, homes addresses and even photographs of scientists on extremist websites has become commonplace. As one scientist expressed, "when you see your name and photograph on a website accusing you of murder and calling for others to 'take (you) out,' it is very disturbing. We now know to how to check for bombs and incendiary devices in or around our offices, homes and vehicles. But that's not something you really want to have to know."

Over the years the level of intimidation has grown. Recently several scientists and others associated with the National Primate Research Centre in Madison have been targeted in their homes. "Home visits," as the animal rights extremists call them, occur at night, oftentimes with the "visitors" masked, hooded or otherwise obscured. One victim of a home visit describes her experience:

> I received an anonymous phone call early in the evening. A male voice asked "Are you home?" He called back and we let the answering machine pick up. This time he said that they were outside and would be there all night. They came up to the house and windows although we had posted no trespassing signs. My husband went outside and told them that they were trespassing. They chanted through a bullhorn and called us out by name stating that we were "f---ing cowards."
>
> My family and I kept a sense of humor and calm about us to set the right tone for the kids. But until you have had strangers dressed in black trespassing onto your front yard and screaming through a loudspeaker and you see your husband doing everything he can to keep them from coming to the front door and frightening the children, and hoping the kids don't hear them call you a "f---ing coward," then you really don't know what it's like. Our house has a lot of windows and sight lines. It's not easy to just go huddle in a room and listen to the loudspeaker blaring,"

The arsons and other attacks against research, education and testing are well-documented. But sometimes more subtle harassments can be devastating as well. A Wisconsin scientist who has endured home "visits," death threats, email barrages and other intimidations has also been the victim of other forms of harassment. She has received hundreds of magazines subscribed for her, including those representing extreme pornography. Other Wisconsin scientists have been victims of vandalism and mischief directed at their property.

Picking up the phone, reading your email or opening a letter or package should not be an act of bravery. Feeling safe in your own home should be a luxury we can afford for our world-leading scientists. Please support the *Animal Enterprise Terrorism Act* and give law enforcement the tools that they need to protect our scientific/animal enterprise.

Thank you.

Gale Davy
Executive Director
Wisconsin Association for Biomedical Research & Education
PO Box 390
Milwaukee, WI 53201-0390
414-899-9246
wabre@execpc.com

PREPARED STATEMENT OF MR. KEITH KAPLAN, EXECUTIVE DIRECTOR, FUR INFORMATION COUNCIL OF AMERICA

FICA

Statement for the Record
on H.R. 4239
"The Animal Enterprise Terrorism Act"

Keith Kaplan
Executive Director
Fur Information Council of America

Before the Subcommittee on Crime, Terrorism and Homeland Security
Committee on the Judiciary
U.S. House of Representatives

The Fur Information Council of America, a national trade association representing 1,550 fur enterprises across the U.S. including fur retailers, manufacturers, merchants, brokers, auctions, agents, dressers and dyers, respectfully submits these comments in support of H.R. 4239, the "Animal Enterprise Terrorism Act." Our industry currently employs approximately 33,000 full-timers and, due to the seasonal nature of our business, nearly 155,000 part-timers. An additional 12,000 or more are employed in related sectors such as transport, insurance, and banking. Retail sales for the industry totaled $1.81 billion in 2004.

For nearly 25 years, animal rights extremists have engaged in direct actions against fur enterprises in virtually every state. Actions have ranged from organized protests to vandalism and, in several cases arson, pipe bombs or Molotov cocktails. Death threats and other forms of intimidation are common. At times, various websites have carried instructions on the construction of incendiary devices along with the addresses of various fur enterprises or even the home addresses of individuals involved in the fur industry.

As animal rights extremists become more savvy and sophisticated in their strategies and tactics, threats against businesses and individuals involved in animal use industries continue to increase and evolve. Between the areas of constitutionally protected advocacy and criminal activity there is a tremendous "grey" area that, in effect, provides an umbrella of protection under the guise of "civil liberties" for the perpetrators of actions that threaten, intimidate and worse.

Further, as the organizational structure of these groups becomes more cunning and cross fertilization of members among these various group increases, it becomes more difficult to counteract their particular brand of terrorism.

Protections granted under the Animal Enterprise Protection Act, passed into law in 1992 are an important tool addressing problems experienced by those involved in certain animal use entities. Amendments in 2002 have sought to expand coverage and increase penalties. However, a significant segment of the population remains unprotected by this statute. Included among this group are fur retailers, manufacturers and designers working with fur, those involved in fur service businesses and even customers of these businesses.

Across the U.S., the animal rights extremists continue to target these businesses and individuals with increasing frequency. Anyone with internet access can easily monitor these actions across a wide range of websites. Protests in front of businesses or homes are common as are phone and fax blockades, gluing of locks, vandalism and other activities designed to disrupt business. In many instances responsibility for these actions is taken or promoted by groups with a national presence such as PETA, ALF, IDA (In Defense of Animals) or WAR (Win Animal Rights) and in photos of these actions we see individuals who have appeared at actions in other states. Animal rights extremists engage in a skillful dance on the ledge of legality, comfortable in the knowledge that either their actions are beyond the reach of prosecutors, or will merely result in a slap on the wrist from state courts. This knowledge empowers them with a sense of "free reign" and has led to the dangerously evolving nature of direct actions, as extremists test the limits of the law in an effort to promote their agendas and achieve their objectives.

Today, highly orchestrated campaigns involve a relentless barrage of threats and vandalism coupled with the promise of escalating tactics or violence. Overall, these campaign have serious economic consequences and create immense personal hardships for the targets and their families. Nevertheless, these activities somehow manage to escape the reach of the law, and therefore, these activities continue unabated. Further, the use of the internet allows for unprecedented effectiveness in the employment of threats and intimidation to businesses and customers alike. The internet also permits the orchestration of campaigns on a national, and even international level.

Consider the recent activities against some of our retail members:

1. In Philadelphia, hundreds of flyers are hung throughout a community and leaflets are distributed at the schools attended by the children of a local furrier featuring his photo and the word "murderer" in big red letters across the page. His home address and phone number are included. At home he receives dozens of threatening phone calls, and he is forced to keep his children home from school for several days due to the harassment they experience.

2. A fur retailer located in a strip mall in Oklahoma City also housing a "market maker" heavily invested in Huntingdon Life Sciences had her windows broken by animal rights extremists so many times that she has been forced to close her business. She can no longer find insurance coverage, her landlord will not renew her lease, and she has spent so much money repairing her windows and increasing her security that she cannot afford to move her establishment nor keep her doors open.

3. The internet is used to cross check license plates of customers or employees seen going to or coming from a fur retailer in Portland, Oregon. Their names, phone numbers and home addresses are posted on animal rights websites with suggestions for direct actions and harassment. Some receive threatening phone calls, some have their automobiles vandalized, others even receive notes threatening their children. At least one of the individuals involved in regular actions against this retailer has been arrested for actions in other states.

4. A retail establishment in San Diego was heavily vandalized as perpetrators enter after breaking windows to destroy computers and merchandise and spray paint defamatory slogans across the walls. Coincidentally, the action occurs in conjunction with an appearance by ELF activist Rodney Coronado at the local university, as well as other direct actions in the community. The owner of the establishment attempts to engage federal law enforcement, advising them that she has video surveillance tapes of the incident. Because of the limited protections provided to her under the Animal Enterprise Protection Act as it currently stands they advise her to work with local law enforcement.

5. An aggressive campaign against a national chain of high end specialty retail stores selling fur involves local actions across the U.S. where extremists chain themselves together with steel sleeves to block entrances during business hours. The campaign is well orchestrated and effectively communicated via the internet. Presently, the activity at any given retail location is addressed on a local level, with few, if any, prosecutions. On a larger scale, animal rights extremists use the potential threat of a similar disturbance of business at hundreds of stores across the U.S. to coerce other multi-unit operators to drop all fur from the merchandise mix.

To be clear, these kinds of activities do not fall under the protection of First Amendment rights. Indeed, our industry is not seeking to invalidate the rights of those who attempt to persuade governments, business or individuals through open debate, discussion or demonstration. Rather, we seek to clearly define those activities that cross the line from free speech to criminal conduct and to assign appropriate guidelines for prosecution and punishment.

Well funded and highly organized, animal rights extremists continue their threat of terrorism against businesses and individuals across the United States. These campaigns of direct action and intimidation cross state lines and even become national in scope, thereby evading the reach of local law enforcement. As these tactics escalate and evolve, the impact on legitimate business is evident. Existing statues do not adequately protect against these evolving activities and, in fact, in their vague language and limitations of coverage, they may actually provide a protective shield under which many of these activities may be carried out.

Prior modifications to the Animal Enterprise Protection Act have provided a base for protecting against violent extremism in the form of direct attacks against animal enterprises. But there is more work to do to eliminate the threat and adequately protect all of our citizens as the extremist tactics have evolved. H.R. 4239 takes that next step by expanding the coverage of the current law to retail establishments; by increasing the penalties for violations under the Act; and most importantly, by providing protections for individuals and enterprises who are subjected to fear and intimidation by these extremists. Moreover, the legislation is carefully crafted to accomplish these objectives without chilling legitimate protest.

We urge Congress to build upon and enhance the legal tools needed to insure the safety and security that are the inalienable rights of every U.S. citizen. Enactment of H.R. 4239 goes a long way toward providing those tools.

PREPARED STATEMENT OF MR. WESLEY J. SMITH, J.D., SENIOR FELLOW, DISCOVERY INSTITUTE

Wesley J. Smith, J.D.
P.O. Box 20958
Castro Valley, CA 94546
e-mail: wjs@wesleyjsmith.com

Testimony of Wesley J. Smith, JD, Before the House Judiciary Subcommittee on Crime, Terrorism, and Homeland Security Hearing: May 23rd, in Support of Strengthening the Law Against "Animal Rights" Terrorism

My name is Wesley J. Smith. I am a lawyer, author, and consumer advocate. I am a senior fellow at the Discovery Institute and a special consultant to the Center for Bioethics and Culture.

For more than ten years I have been deeply engaged internationally in public policy debates about the most important bioethical issues facing our nation and global community. These include researching and writing about the dangers of the radical animal rights/liberation movement. I am in the midst of researching a book that I plan to write on this subject. It will be my 12th published book. My abridged CV is attached to my testimony.

My work in the fields about which I advocate is entirely secular, which I believe is appropriate to the creation of public policy in a nation governed by the rule of law.

I am submitting this written testimony in support of laws that would increase the legal protection for both people who work in animal industries as well as those who work in ancillary businesses that contract or otherwise have business dealings with such companies. The latter category of protection is important because of a tactic now being utilized by animal rights terrorists known as "tertiary targeting."

Tertiary targeting is a profoundly troubling and brutally effective method of attacking animal using "target" businesses. Knowing that every modern company relies on other businesses for services like banking, insurance, and auditing, perpetrators of tertiary targeting harass and intimidate the executives, workers (and even their spouses and children) of companies doing business with targets companies

Tertiary targeting has nothing in common with peaceful picketing or even civil disobedience. The most vicious example of tertiary targeting has been the international terrorist campaign directed against an animal testing company called Huntingdon Life Sciences by a loose organization of liberationists called "Stop Huntingdon Animal Cruelty" (SHAC). Here's roughly how tertiary targeting works: The SHAC WEB site identifies targets to be harassed and provides information for use by harrassers, including home addresses, phone numbers, names and ages of their children and even where they attend school. Activists send anonymous death threats, mail targets video tapes of their family members, and vandalize their property.

As of May 2006, a SHAC WEB site listed scores of companies for targeting, including: Abbott Labs (U.S. based with sites throughout the world), American Pacific Corporation (U.S.-based chemical company), Bristol Myers Squibb (U.S.-based international pharmaceutical corporation), Kumiai Chemical Industry Co. (Japan-based herbicide company), and Merck (U.S.-based pharmaceutical company).

To avoid being victimized or to surrender and be removed from the target list, companies are told on SHAC's WEB site (http://www.shac.net/TARGETS/suppliers/suppliers.html):

> TO ALL SUPPLIERS: If you have severed your links with Huntingdon Life Sciences, please let the campaign know. You can send a simple email to info@shac.net stating the following: "......(name of your company) have severed their links with HLS and terminated their contract, and will not be dealing with them now or in the future, directly or

> indirectly." This will enable supporters to be kept up to date with which companies are still involved with Huntingdon Life Sciences.

Failing to do so opens companies to the threat of tertiary targeting.

This is terrorism, pure and simple—and unfortunately, it is working. SHAC and its allies like the Animal Liberation Front (ALF) have scared more than one hundred businesses into cutting off ties with Life Sciences, including the huge auditing firm of Deloitte & Touche, with the increasing numbers of notches on SHAC's gun then posted proudly on its WEB sites. At present, SHAC lists 113 companies on its WEB site that have pledged to comply with SHAC's demands business with Life Sciences including the international corporations Johnson and Johnson, Washington Mutual, UBS Global Capital, Nucryst Pharmaceutical, and Chubb.

In October 2005, the New York Stock Exchange planned to Huntingdon Life Sciences (renamed Life Sciences Research) on the Big Board. This was a very big deal for Life Sciences. It had previously been de-listed by the Exchange because attacks on the company by animal liberationists had undermined its financial stability. But then, on the very day the company was to be eligible for trading, the Exchange rescinded the listing. Executives refused to explain their decision—even to a United States Senate committee—but it was surely more than coincidental that the rescission came immediately after liberationists vandalized an executive's yacht club and threatened to target Exchange employees for attack.

Other violent animal liberationist groups also have contributed to a crescendo of violence and intimidation against lawful businesses in recent years. For example, an instruction manual posted on the ALF WEB site entitled, *Arson-Around With Auntie ALF: Your Guide to Putting Heat on Animal Abusers Everywhere*, teaches readers how to form a self-contained terrorist cell, commit arson, and then claim responsibility for the crime on behalf of ALF. Since only the perpetrators will know who set the fire, the manual asserts, the chances of being caught or penetrated by law enforcement are close to zero.

To be sure, most liberationists do not personally engage in such threats and violence. But very few leaders (or grass roots activists) condemn these tactics. A few even openly support it. For example, Bruce Friedrich, the second in command at People for the Ethical Treatment of Animals (PETA), told an animal liberation conference in 2001:

> Of course, we're going to be as a movement blowing stuff up and smashing windows. For the record, I don't do this stuff, but I do advocate it. I think it is a great way to bring about animal liberation. And considering the level of the atrocity and the level of the suffering, I think it would be a great thing if all of these fast food outlets and slaughter houses and laboratories—and the banks that fund them—exploded tomorrow."

The Ideology of Animal Liberation

To understand why people could become so committed to the liberationist cause that they would threaten children and even rob the graves of targeted families (as happened recently in the United Kingdom), it must be understood that the ideologues of animal liberation do not believe in the same moral principles as do committee members, and your constituents.

To a true believer in animal rights, human life does not have special value. Rather, what confers moral value on an organism is the ability to feel pain. Thus, animal liberationist Professor Richard Ryder, formerly of Tulane University, wrote that "painience"—his term for the ability to feel pain—"is the only convincing basis for attributing rights, or indeed, interests to others." In this view, since most animals can feel pain, "non-humans" belong in "the same moral and legal circle as ourselves," meaning that humans should not "be able to exploit them as slaves."

This belief means that to animal liberationists, whatever is done to an animal *is the same as if it were being done with and to a human.* This is not metaphorical or a symbolic assertion: Adherents *believe literally* that cattle ranching is morally equivalent to human slavery and that animal research is torture of the kind experienced by the inmates of the concentration camps. Dr. Steven Best, a University of Texas philosophy professor and international star of the animal rights movement, argues that normal practices of animal husbandry are akin to the attacks on the World Trade Center and Pentagon on September 11, 2001: "The hellish reality of animal existence cannot fundamentally change until we create a seismic cultural shift that replaces the notion of animals as property with a radically alternative concept, such as animals as persons." To Best, the slaughter of farm animals and the demise of animals in medical testing is akin to terrorism. Hence, "For the animals, every second is a 9/11 attack." Perhaps not coincidentally, Best was barred from the United Kingdom in 2005 for supporting lawlessness in the name of animal rights.

Animal rights/liberation must be distinguished from "animal welfare." The former is dedicated to preventing *any* use of animals by humans—whether as food, in research, or even as seeing-eye dogs—regardless of the substantial human harm thereby caused. In contrast, the animal welfare movement acknowledges human uniqueness and recognizes man's duty to treat animals humanely and not cause them gratuitous pain or suffering. (Thus PETA is animal liberationist, while the Society for the Prevention of Cruelty to Animals is a notable animal welfare organization.)

The differences between these contrasting approaches to animal protection could not be starker. For example, is the life of a monkey as precious as that of a human being? Animal rights believers say, yes. Welfarists say, no. Is butchering a cow the moral equivalent of lynching a black man during the Jim Crow era? Welfarists would scoff. But PETA's "Animal Liberation Project" stated explicitly that the two are morally equivalent. Are the lives of people more important than the lives of animals? Not according to PETA: "We are all animals," the groups asserted in its Animal Liberation Project, by which it was not stating a biological fact but asserting an explicit human/animal moral equality. Welfarists, on the other hand, cringe at the comparison between Apartheid-style bigotry and proper animal husbandry.

It is the radical and subversive ideology of animal rights/liberation justifies (in some minds) lawlessness and violence—just as many would accept violence to prevent the trains from reaching Auschwitz.

A few liberationists even contemplate the potential of resorting to murder to protect the animals. In December 1998 animal rights terrorists threatened to murder ten identified medical researchers. After Prime Minister Tony Blair refused to create a commission to investigate alleged abuses in animal research, a convicted "animal rights" felon named Barry Horne— in prison for torching a department store because it sold fur coats—began a hunger strike. ALF then issued a deadly threat on behalf of another terrorist group, the Animal Rights Militia (ARM): "The ARM has announced a list of ten vivisectors who will be assassinated if animal liberation hunger striker Barry Horne dies through Labour's broken promises." (Horne subsequently called off his strike and the crisis passed.)

In the United States, militant Jerry Vlasak justified the murder of scientists who research with animals at a 2003 animal rights convention. "I don't think you'd have to kill—assassinate too many [scientists]," he declared. "I think 5 lives, 10 lives, 15 human lives could save a million, 2 million, 10 million non-human lives." He later told a stunned United States Senate Committee that animals and humans are "morally equal," asserting that the "murder" of those "who hurt animals and will not stop after being told to stop" is "morally justified."

The first "animal rights murder" may already have taken place. In the Netherlands, an animal rights extremist assassinated a popular candidate for parliament, perhaps because he defended pig farming in a debate with animal rights activists.

Conclusion

The federal government has an urgent role in protecting the proper and humane use of animals. Understanding why extremists in the animal rights/liberation movement are resorting to terrorist tactics is necessary in order to understand that strong action is urgently needed. It is my hope that this brief overview has been helpful in this regard and I stand ready to assist the Committee in any way that I can.

Wesley J. Smith,
Senior Fellow
Discovery Institute,
Seattle, WA

WESLEY J. SMITH

Award winning author and lawyer Wesley J. Smith is a senior fellow at the Discovery Institute, an attorney for the International Task Force on Euthanasia and Assisted Suicide, and a special consultant for the Center for Bioethics and Culture. In May 2004, because of his work in bioethics, he was named by the *National Journal* as one of the nation's top expert thinkers in bioengineering.

Smith left the full time practice of law in 1985 to pursue a career in writing and public advocacy. He has authored or co-authored eleven books. He formerly collaborated with Ralph Nader, co-authoring four books with consumer advocate: *Winning the Insurance Game* (1990), *The Frugal Shopper* (1991), *Collision Course: The Truth About Airline Safety* (1993) and *No Contest: Corporate Lawyers and the Perversion of Justice in America* (1996). He also co-authored (with Eric M. Chevlen, MD), *Power Over Pain*, a consumer's guide to obtaining good pain control.

His book *Forced Exit: The Slippery Slope from Assisted Suicide to Legalized Murder* (1997), a broad-based criticism of the assisted suicide/euthanasia movement, has become a classic in anti-euthanasia advocacy. First published in 1997 and in revised paperback in 2003, it was updated again in 2006 and published by Encounter Books under the new title: *Forced Exit: Euthanasia, Assisted Suicide, and the New Duty to Die.*

Smith's *Culture of Death: The Assault on Medical Ethics in America*, a warning about the dangers of the modern bioethics movement, was named one of the Ten Outstanding Books of the Year and Best Health Book of the Year for 2001 (Independent Publisher Book Awards). Smith's *Consumer's Guide to a Brave New World* (2005), explores the morality, science, and business aspects of human cloning, stem cell research, and genetic engineering. He is also conducting research for a planned book about the animal liberation movement.

Smith's writing and opinion columns on assisted suicide, bioethics, the morality of human cloning, the dangers of animal liberation, legal ethics, and public affairs have appeared nationally and internationally, including in *Newsweek, New York Times, The Wall Street Journal, USA Today, Forbes, The Weekly Standard, National Review, First Things, The Age* (Australia), *Western Journal of Medicine*, and the *American Journal of Bioethics*. He has also been published repeatedly in regional publications including in the *San Francisco Chronicle*, the *Seattle Times*, the *Dallas Morning News*, the *Rocky Mountain News*, the *Detroit News*, the *Orange County Register*, the *New York Post*, and many other newspapers throughout the nation. He is also a frequent source for journalists reporting on issues about which Smith advocates.

Throughout the course of his career in public advocacy, Smith has appeared on thousands of television and radio talk/interview programs, including on such national programs as ABC Nightline, Good Morning America, Larry King Live, CNN Crossfire, CNN World

Report, the CBS Evening News, Coast to Coast, the Dennis Prager syndicated radio show, the Mike Gallagher syndicated radio show, the Tony Snow radio show, Afternoons with Al Kresta, EWTN, CSPAN-Book TV, Fox News, and CNN Talk Back Live. He has also appeared internationally on Voice of America, CNN International, and on programs originating in Great Britain (BBC), Australia (ABC), Canada (CBC), New Zealand, Germany, China, and Mexico.

Smith is an international lecturer and public speaker, appearing frequently at political, university, medical, legal, disability rights, bioethics, religious, and community gatherings across the United States, Europe, Canada, South Africa, Mexico, and Australia.

PREPARED STATEMENT OF MR. JAMES C. GREENWOOD, PRESIDENT AND CEO, BIOTECHNOLOGY INDUSTRY ORGANIZATION

Bio
BIOTECHNOLOGY
INDUSTRY ORGANIZATION

Testimony
James C. Greenwood
President and CEO
Biotechnology Industry Organization

Subcommittee on Crime, Terrorism, and Homeland Security
Committee on the Judiciary
U. S. House of Representatives

Legislative Hearing on H. R. 4239, the "Animal Enterprise Terrorism Act "

May 23, 2006

Chairman Coble, Ranking Member Scott, and Members of the Crime, Terrorism, and Homeland Security Subcommittee,

Thank you for providing the opportunity to submit written testimony on H. R. 4239, the "Animal Enterprise Terrorism Act." BIO supports the basic intent and general provisions of the Animal Enterprise Terrorism Act.

BIO is an industry organization whose members expand the boundaries of science to benefit mankind by providing better healthcare, enhanced agriculture, and a cleaner and safer environment. The mission of BIO is to be the champion of biotechnology and the advocate for its member organizations –both large and small.

BIO represents more than 1,100 biotechnology companies, academic institutions, state biotechnology centers and related organizations in 50 U.S. states and 31 other nations. BIO members are involved in the research and development of health care, agricultural, industrial, and environmental biotechnology products.

As the leader of the world's largest biotechnology association, I am reminded everyday that biotechnology promises to transform the world. That promise, however, is increasingly threatened by animal rights extremists. Our industry must be able to conduct research and development with due protection under the law from illegal acts and harassment routinely committed by certain individuals associated with the animal rights movement. My testimony focuses on our support for the Animal Enterprise Terrorism Act, the important and necessary role of animals in biotechnology research, BIO's ethical principles of the humane care and use of animals in research, recent assaults on BIO members by certain animal right extremists, and the importance of protecting the biotechnology industry by passing the Animal Enterprise Terrorism Act.

Attacks on the Biotechnology Industry

For more than a decade, several biotechnology and pharmaceutical companies have been attacked by animal rights extremists and eco-terrorists, both domestically and internationally. Animal rights extremists have engaged in bombings of research facilities, harassment of the spouses, children, friends and neighbors of biotech executives, and vandalism of personal property. They have extended this terrorism to include 'tertiary targeting' or 'third party targeting'. Their campaign against biotechnology companies is strategic, specific, unrelenting and directed toward delivering economic and physical damage to companies engaged in research for cures to diseases ranging from cancer to cystic fibrosis. These campaigns have cost targeted biotechnology companies millions of dollars to repair the vandalism, to add extraordinary protections for their employees and business systems and to seek legal protections from the courts to assure that operations can continue. These resources would have been more appropriately spent on investment in clinical research to find cures and preventatives for catastrophic and debilitating diseases plaguing millions of Americans. In addition, farms, ranches, processors and retailers of animal-based food products have been targeted by these groups. The FBI and the Bureau of Alcohol, Tobacco and Firearms have formally classified eco- and animal rights terrorism as this country's most serious domestic terrorist threat.

Importance of the Role of Animals in Research

Ethical animal research has played a vital role in virtually every major medical advance of the last century – for both human and animal health. This research is invaluable in the development of life-extending treatments for people, as well as cats, dogs, farm animals, wildlife and endangered species. The ability to conduct humane and responsible animal-based research must be preserved to help conquer disease, alleviate suffering, and improve the quality of life. Biotechnology companies have depended on this research to develop more than 200 drugs and vaccines approved by FDA, helping 325 million people worldwide and preventing incalculable human suffering.

BIO members are compelled by ethical concerns and legal requirements to evaluate the safety and efficacy of potential medicines and food products before they are given to humans and animals; the use of animals in research is a requirement for many such products. The appropriate and responsible use of animals is therefore an indispensable part of biotechnology research that includes biomedical and agricultural research.

A Priority of the Biotechnology Industry - Humane Care and Use of Animals

BIO members are committed to improving the quality of human and animal life with biotechnology through humane and responsible animal-based research. BIO believes that such use is a privilege, imposing a responsibility to provide proper care and humane treatment in accordance with the BIO Statement of Ethical Principles for the Care and Use of Animals in Biotechnology Research.

Support for the Animal Enterprise Terrorism Act

Because of the important role of animals in biotechnology research and our commitment to their humane care and use, BIO supports the basic intent and general provisions of the Animal Enterprise Terrorism Act, H. R. 4239, and its companion bill in the Senate, S. 1926. This legislation would strengthen and expand federal criminal sanctions against animal rights extremists and eco-terrorists. The bill would close existing loopholes in federal law that currently permit these extremists to intimidate and harass individual executives and researchers, or target for disruption or vandalism, property of those affiliated with animal enterprises, without fear of federal prosecution.

The bill would provide effective protection to the biotechnology industry, including our human health, food and agriculture sectors. 'Animal enterprise' is defined to include a commercial or academic enterprise that uses or sells animals or animal products for profit, food or fiber production, agriculture, research, or testing.

Although our industry currently conducts business under the protection of the Animal Enterprise Protection Act of 1992 (18 USC Sec. 43), this new legislation is needed to enhance the effectiveness of the Department of Justice's response to recent trends in the animal rights extremist movement. The bill addresses the "tertiary targeting" or "third party targeting" system used by animal rights extremists, by prohibiting the intentional damaging of property of a person or entity having a connection to, relationship with, or transactions with an animal enterprise. Currently, only the animal enterprise itself is protected by existing law.

The legislation also makes it a federal crime to intentionally cause a person to be in reasonable fear of death or bodily injury (to such person or the immediate family of such person) through threats, vandalism, property damage, trespassing, harassment, or intimidation because of their involvement in an animal enterprise. In addition, it increases penalties for intentionally causing economic disruption or damage.

I have stated previously that BIO has zero tolerance for animal rights extremism. On September 7, 2005, when the New York Stock Exchange (NYSE) postponed the listing of Life Sciences Research (Huntingdon Life Science) in an apparent reaction to threats from animal rights terrorists, I sent a strongly worded letter to John Thain, NYSE chairman, and Catherine Kinney, NYSE president, asking them to reconsider their decision. In that letter, I stated that the postponement of this listing suggests that biomedical research has taken a backseat to the pressure tactics of extremists. BIO companies stand together in our opposition to any efforts that would responsibly hinder our ability to develop and produce products that are safe for patients, consumers, populations, pets, livestock, wildlife, and ecosystems. To date, no response has been received by BIO from the NYSE.

Recent results of the federal criminal trial in New Jersey involving the members of Stop Huntingdon Animal Cruelty (SHAC) is a victory for the tens of thousands of researchers, scientists, and other employees working in the biotechnology and pharmaceutical industries. Our industry applauds the jury verdict holding SHAC accountable for conspiracy to commit terrorism and other unlawful acts. The United States has a proud tradition of navigating the line between free speech and criminal conspiracy, which the defendants in that case clearly crossed. This prosecution and verdict is a strong signal to those who would attempt to intimidate researchers and their families, friends and business associates. The criminal actions of SHAC and other animal rights extremists are **not** protected by the First Amendment, and as a society, we cannot allow the minority view of certain extremists to hold hostage our collective efforts to improve the lives of humanity.

The future ability of the biotechnology industry to improve the quality of human and animal life rests on our ability to conduct research that leads to the approval of new products. Humane animal research is an essential part of delivering these life-saving products to humankind. We must provide the Department of Justice the necessary authority to apprehend, prosecute, and convict individuals committing animal enterprise terror. BIO urges passage of H.R. 4239.

Prepared Statement of the California Healthcare Institute (CHI)

CHI California Healthcare Institute www.chi.org

Statement Of The
California Healthcare Institute

Submitted to the
Committee on the Judiciary
Subcommittee on Crime, Terrorism, and Homeland Security
Of The
United States House of Representatives

Legislative Hearing On
H.R. 4239, the Animal Enterprise Terrorism Act
May 23, 2006

The California Healthcare Institute (CHI) appreciates the opportunity to submit written testimony on H.R. 4239, the Animal Enterprise Terrorism Act, and thanks Chairman Coble, Ranking Member Scott, and Members of the Crime, Terrorism, and Homeland Security Subcommittee for their attention to this important matter.

CHI represents more than 270 of California's leading biotechnology, pharmaceutical, and medical device companies as well as our state's premier academic and non-profit research institutions. The California life sciences community employs nearly 250,000 workers, provides over $14 billion in wages and salaries, and accounts for over $7 billion in global exports. The state's biomedical industry invests some $15.5 billion annually in researching and developing innovative therapies and devices, with the average company investing 48 percent of its revenues back into research and development (R&D). And California universities, medical schools, and research institutions are global leaders in basic and clinical research – together receiving more funds from the National Institutes of Health (NIH) than any other state.

Threat to medical research/innovation posed by criminal conduct of animal rights extremists

Over the past number of years, California-based biomedical companies and research institutions have increasingly been the target of a persistent terrorist campaign. With alarming consistency, members of Stop Huntingdon Animal Cruelty USA (SHAC USA), the Animal Liberation Front (ALF), and other groups have subjected researchers, company executives and their children, and employees to personal property destruction, harassment, and threats.

Importantly, the targets of this violence have never violated, nor been accused of violating, any federal regulation or statute in the care and use of animals.

In March 2005, John E. Lewis, Deputy Assistant Director, Counterterrorism Division, Federal Bureau of Investigation, made the threat clear stating, "In recent years, animal rights extremists

Headquarters
1020 Prospect Street, Suite 310
La Jolla, California 92037
858.551.6677 ▪ Fax 858.551.6688

Sacramento
1215 K Street, Suite 970
Sacramento, California 95814
916.233.3497 ▪ Fax 916.233.3498

and eco-terrorists have become the most active criminal extremist elements in the United States."[1]

Unfortunately, current provisions in the Criminal Code resulting from the enactment of the 1992 Animal Enterprise Protection Act and its amendments in 1996 and 2002, fail to provide law enforcement with effective tools to use against these terrorists.

It is therefore vitally important that the Congress take action to enable the prosecution of these violent protestors, whose perverse actions threaten to chill future research and innovation towards cures for cancer, Alzheimer's, AIDS, and other ailments.

That is why CHI supports passage of H.R. 4239, the Animal Enterprise Terrorism Act.

An overview of incidents in California

A cursory review of press reports and other publicly available information paints a disturbing picture of the tactics of these extremist groups in California in recent years. The following are but a sample:

> **March 2003 -August, 2004** – SHAC activists waged an extensive campaign targeting Chiron Corporation, as well as its executives and their children during a 17-month long campaign. The activists directly attacked the company's communication systems; harassed employees in their homes, in one case breaking and entering homes and destroying personal property; executives were subjected to photographic surveillance by the activist and their local neighborhoods were the site of demonstrations by masked activists; and activists verbally harassed and engaged in intimidation tactics several of the children of employees during soccer games and while at school. In August 2003, a domestic terrorist group, identified as The Revolutionary Cells, planted two pipe bombs at Chiron Corporation's headquarters in Emeryville, CA. The bombs were timed to explode several minutes apart, in order to injure or kill firefighters and public safety workers as they responded to the first blast.
>
> **March, 2003** -- ALF activists shoot out the glass front door of the Beverley Hills, CA offices of E-Trade, which trades Huntingdon Life Sciences.
>
> **April, 2004** -- UCLA Primate Freedom Project and student group protest and vandalize the home of a UCLA researcher.
>
> **November 2004** - ALF and SHAC activists leave threatening voice mails for senior Allergan, Inc executive; Company website records further threats from ALF and SHAC activists.

---

[1] John E. Lewis, Deputy Assistant Director, Counterterrorism Division, Federal Bureau of Investigation, 4th Annual International Conference on Public Safety: Technology and Counterterrorism, Counterterrorism Initiatives and Partnerships, San Francisco, California, March 14, 2005

July 2005 --An ALF video posted on the Internet shows hooded activists spray-painting slogans and smashing windows at Bachem Laboratories in San Carlos, CA. A warning posted on the Net said '"Bachem -- this is only the beginning. All of the security, cameras and burglar alarms that you can buy will not save you now. Sever any and all ties to HLS. If not, things can only get worse. See you soon!"'

September 2005 -- ALF activists attack the home of the Los Angeles Animal Services GM, Guerdon Stuckey. The claim on ALF's website says "The animal liberation front has taken advice from our Commander-in-Chief to 'smoke terrorists out of their holes.' The target was Los Angeles Number One terrorist Gordon Stuckey. Military Strength Smoke grenades were detonated on the floor of this animal killer's abode. Sleep light...alf." ALF also claimed responsibility for a bomb threat to LAAS Commander Dave Diliberto: "The message was left on David's cell phone stating a bomb had been left in his car, the car paid for with the blood of innocent animals. Consequently, his office, the neighborhood, had to be evacuated."

To better determine the extent of these threats in the state, CHI recently developed a confidential survey of California biomedical firms, research institutions, and universities in order to quantify the volume of incidents and threats perpetrated by these groups in the state. The preliminary findings of this survey are attached and further illustrate, we believe, the extent to which these activities go far beyond that which could be defended as legitimate and protected First Amendment rights exercises.

Conclusion

Biomedical research and development in California is under serious threat from extremists acting in the name of animal rights. Company employees, researchers, and their families continue to be the subject of ongoing campaigns of intimidation on a scale unforeseen when the original Animal Enterprise Protection Act and subsequent amendments were enacted.

Indeed, the tactics of radical activists have evolved in ways not anticipated by existing law. New tactics include a coordinated international campaign against third-party companies that support services to companies and academic research institutions that conduct animal research. Targets include banks, insurance companies, stockbrokers, customers, construction services, food services, Internet service providers, telecom companies, even janitorial services.

Therefore, to address the threat posed by terrorist acts committed against research laboratories, businesses, and other entities, and to safeguard continuing biomedical research into the treatments, therapies, and medical technologies of the future, CHI urges passage of H.R. 4239, the Animal Enterprise Terrorism Act.

Thank you.

RESULTS OF PRELIMINARY SURVEY OF THREAT POSED BY ANIMAL RIGHTS EXTREMISTS, CONDUCTED BY THE CALIFORNIA HEALTHCARE INSTITUTE (CHI)

C H I California Healthcare Institute
www.chi.org

Threat to medical research/innovation posed by criminal conduct of animal rights extremists

Over the past number of years, California-based biomedical companies and research institutions have increasingly been the target of a persistent terrorist campaign. With alarming consistency, members of Stop Huntingdon Animal Cruelty USA (SHAC USA), the Animal Liberation Front (ALF), and other groups have subjected researchers, company executives and their children, and employees to personal property destruction, harassment, and threats. Importantly, the targets of this violence have never violated, nor been accused of violating, any federal regulation or statute in the care and use of animals. In March 2005, John E. Lewis, Deputy Assistant Director, Counterterrorism Division, Federal Bureau of Investigation, made the threat clear stating, "In recent years, animal rights extremists and eco-terrorists have become the most active criminal extremist elements in the United States."

Unfortunately, current provisions in the Criminal Code resulting from the enactment of the 1992 Animal Enterprise Protection Act and its amendments in 1996 and 2002, coupled with recent U.S. Supreme Court rulings eroding the useful application of the Racketeer Influenced and Corrupt Organizations (RICO) Act, fail to provide law enforcement with effective tools to use against these terrorists. It is therefore vitally important that the Congress take action to enable the prosecution of these violent protestors, whose perverse actions threaten to chill future research and innovation towards cures for cancer, Alzheimer's, AIDS, and other ailments.

To determine the extent of these threats in the state, CHI developed a confidential survey of California biomedical firms, research institutions, and universities in order to quantify the volume of incidents and threats perpetrated by these groups in the state. The preliminary findings of this survey are included below.

PRELIMINARY SURVEY FINDINGS – February 2006

Respondents

Companies = 11	Bay Area = 9
Universities = 4	Sacramento = 1
Research Institute = 2	Los Angeles = 1
	Orange County = 2
	San Diego = 4

Organizations targeted, past 12 months – 53%

Threats/Attacks targeted...

- Facility – 70%
- Individual Employee – 80%
- Employee's Family Member – 30%

If targeted, incidents reported to...

Local law enforcement – 80%
FBI – 60%

Were any arrests made?

Yes –25%
No – 50%
Don't Know – 25%

Incidents involved...

- Threat of violence to employee – 60%
- Threat of violence to family member – 50%
- Threat of physical damage to company property – 40%
- Threat of physical damage to personal property – 30%
- Actual violence to employee – 10%
- Actual physical damage to company property –20%
- Actual physical damage to personal property – 20%
- Distributing leaflets in neighborhood, etc. – 70%

Have you taken measures to enhance security to respond to threat of attack, *regardless* of whether you have been yet targeted?

YES – 88%
NO -- 12%

Have there been significant additional costs associated with adopting such measures?

YES –93%
NO -- 7%

HEADQUARTERS
1020 Prospect Street, Suite 310
La Jolla, California 92037
858.551.6677 ▪ Fax 858.551.6688

SACRAMENTO
980 Ninth Street, Suite 2200
Sacramento, California 95814
916.498.3307 ▪ Fax 916.498.3372

LETTER TO THE HONORABLE HOWARD COBLE FROM THE ANIMAL ENTERPRISE PROTECTION COALITION (AEPC)

AEPC
ANIMAL ENTERPRISE PROTECTION COALITION

May 22, 2006

The Honorable Howard Coble
Chairman
House Judiciary Subcommittee on
Crime, Terrorism, and Homeland Security
207 Canon House Office Building
Washington, D.C. 20515

Dear Chairman Coble:

On behalf of the undersigned member organizations of the Animal Enterprise Protection Coalition (AEPC), we are writing to express our group support for H.R. 4239 and S.1926, the Animal Enterprise Terrorism Act (AETA) of 2005. We strongly believe that current federal law should be amended to address the very serious threat of violent actions committed by animal rights extremists pose to scientific research.

The FBI considers animal rights extremists among its most serious domestic threats. Their cause is typically advanced through "direct action" which includes death threats, vandalism, and bombings. In addition, they calculate to aggressively intimidate and harass individuals who are identified as targets. While traditional targets have included research and biomedical laboratories, the Internet and other technological advances have provided extremists with additional ways to terrorize an increasing number of people. These groups are now taking direct action against third-parties. Third parties, or tertiary targets, are individuals who are related to a targeted researcher or companies who do business with a research facility. Recent victims have included: families, neighbors, friends, co-workers, individuals delivering goods and services, and other businesses carrying out contractual obligations to the researcher or animal enterprise.

In recent congressional testimony, a Justice Department official said that, "…SHAC [Stop Huntingdon Animal Cruelty] and other animal rights extremists have recognized limits and ambiguities in the [current] statute and have tailored their campaign to exploit them." SHAC is a UK-based extremist group which has taken responsibility for violent actions in the UK and the US. Recently, six SHAC extremists were convicted by a federal jury in Trenton, New Jersey on charges of telephone harassment, interstate stalking, and conspiracy.

The Animal Enterprise Terrorism Act would address gaps in the law that keep authorities from using it in the most effective manner possible. Drafted with technical assistance from the counter-terror experts at the Department of Justice and Federal Bureau of Investigation, the bill provides penalties for intentional economic damage and disruption, and for intentionally causing bodily harm or placing a person in reasonable fear of death or bodily harm. It also specifically addresses "tertiary targeting" or third-party targeting tactics by prohibiting intentional damage of property belonging to a person or organization with ties to an animal enterprise. The AETA will allow federal authorities to help prevent, better investigate, and prosecute individuals who seek to halt medical research through acts of intimidation, harassment and violence.

I urge you to support the Animal Enterprise Terrorism Act of 2005 by becoming a co-sponsor.

Sincerely,

AEPC
ANIMAL ENTERPRISE PROTECTION COALITION

American Academy of Sleep Medicine (AASM)
American Association for Laboratory Animal Science (AALAS)
American Brain Coalition (ABC)
Americans for Medical Progress (AMP)
American Physiological Society
American Psychological Association (APA)
American Society of Laboratory Animal Practitioners (ASLAP)
American Society for Pharmacology & Experimental Therapeutics (ASPET)
Animal Agriculture Alliance
Association of American Medical Colleges (AAMC)
Association of American Veterinary Medical Colleges (AAVMC)
Association for Assessment and Accreditation of Laboratory Animal Care International (AAALAC)
Association of American Universities (AAU)
Association for Psychological Science (APS)

Boehringer Ingelheim Pharmaceuticals
Boehringer Ingelheim Vetmedica, Inc.
Biotechnology Industry Organization (Bio)

California Biomedical Research Association
Cedars-Sinai Health System
Covance

Farm Animal Welfare Coalition
Federation of American Societies for Experimental Biology (FASEB)

GlaxoSmithKline

Human Nutrition Research Center on Aging at Tufts
Huntingdon Life Sciences (HLS)

Massachusetts Society for Medical Research (MSMR)
Michigan Society for Medical Research (MISMR)
Mid-Continent Association for Agriculture, Biomedical Research and Education (MAABRE)

National Animal Interest Alliance (NAIA)

New Jersey Association for Biomedical Research (NJABR)
Northwest Association for Biomedical Research (NWABR)

Ohio Scientific Education & Research Association (OSERA)

Pennsylvania Society for Biomedical Research (PSBR)
Pet Industry Joint Advisory Council (PIJAC)
Pfizer

Sleep Research Society
Society for Experimental Biology and Medicine (SEBM)
Society for Neuroscience (SfN)
States United for Biomedical Research (SUBR)
Southern Research Institute (SRI)

Texas Society for Biomedical Research (TSBR)

University of Texas Medical Branch

LETTER FROM F2 CHEMICALS LTD

F2 Chemicals Ltd
Lea Lane
Lea Town
Preston
Lancashire
PR4 0RZ
UK

25 May 2006

To Whom It May Concern,

I am the Managing Director of a small speciality chemical company based in Preston, Lancashire, United Kingdom called F2 Chemicals Ltd. The Company employs about 40 people and has an annual turnover of £5m and is a subsidiary of a Japanese company called Asahi Glass Co Ltd.

Asahi Glass Co Ltd was identified by Animal Rights Activists as a customer of Huntingdon Life Sciences (HLS), one of the leading animal testing facilities in Europe. As a result, three UK subsidiaries of Asahi Glass Co Ltd, including F2 Chemicals Ltd, became targets of Stop Huntingdon Animal Cruelty (SHAC) as part of its objective to close down HLS.

Since April 2003 the three companies, their employees and directors have collectively been subjected to the following:-

- More than 10,000 e-mails (some threatening and abusive)
- More than 5,000 nuisance telephone call (some threatening and abusive)
- More than 1,000 letters (some threatening and abusive)
- 4 hoax parcel bombs
- 5 people were alleged to be convicted paedophiles
- approximately 35 factory demonstrations
- more than 30 "home visits"

Examples of threatening e-mails: "Animal testing is evil. I would happily go to prison just for the satisfaction of stabbing an animal tester to death. I'll find you and fuck you up. Your company is fucked. Unless you stop dealing with Huntingdon Life Sciences and testing on animals", "........this won't be over until I see your faces smashed against the bottom of your coffins"

Example of a threatening letter that I received at home: "You support animal cruelty and torture at Huntingdon life sciences. You are a sick evil pervert. You are now the ALF priority target. Your life is about to become a living hell until you stop abusing animals for profit. You vile filthy deviant". (ALF stands for Animal Liberation Front)

To put some of the following information into perspective it is important to know that I have a wife and in 2003 our daughter was 6-year old and we live in a remote rural location.

On Saturday 5 July 2003 I received a hoax parcel bomb at my home. The police and the army bomb disposal people were called to deal with the situation. The army captain described the package as "having the appearance of a classic parcel bomb". The police and/or army personnel were present at my home for over 6-hours.

Around 4 August 2003 several hundred letters were sent to people within a 5-mile radius of my home, these letters gave my name and address and stated that I was a convicted paedophile.

We have also experienced 13 "home visits". During these visits the activities of the activists have ranged from making very loud noise using sirens or fireworks during the early hours of the morning to criminal damage. As an example, at about 03:30 on Wednesday 20 August 2003 four extremists were recorded on CCTV after spray painting my home and two cars. On another occasion 14 statements were spray painted over about ½ mile of roadway near my home. Examples of the statements made are as follows, "BILL THE A.L.F. ARE COMING FOR YOU!". "BILL THE MURDERER" "WE KNOW WHERE YOU LIVE"

As a result of the above activities my wife was off work ill with stress for over 2 months. To try to protect my family my shareholder employed a specialist security firm and security guards were present at my home around the clock for about 9-months.

A Director of a sister company and his wife suffered similar harassment to myself and my family. They also had security personnel at their home for about 1-year.

The activists then started to target other employees of the companies. My Quality Assurance Manager and his wife had windows broken on their two cars at 00:30 hours on Tuesday 13 January 2004. A shift process worker at a sister company had paint stripper dribbled on his car and graffiti sprayed on the front of his house.

My Company, in conjunction with a number of other Japanese companies, obtained a UK High Court Injunction under the 1997 Protection for Harassment Act. This was partially successful in causing a reduction to the problem.

The total cost to Asahi Glass for increased security at its factories and for domestic security, legal fees and damage rectification was over £1m. Non-financial impacts include frightened and/or stressed employees and their friends, relatives and neighbours. A number of staff have left employment and others have considered leaving. A huge amount of management and staff time and focus has been redirected to counter the threat.

In 2004 I became a founder member of an organisation called "Victim's of Animal Rights Extremism" (VARE). This organisation was formed with support from the UK's major political parties with the twin objectives of helping other victims of animal rights extremism and lobbying the UK Government to solve the problem. VARE worked successfully with a number of other organisations to get additional legislation enacted to assist in the fight against the extremists. VARE now works closely with the UK police and other governmental organisations to improve awareness of the impact of this type of extremism.

It is clear to me, from information that I have been given from police and other sources over the last few years, that the animal extremism problem is on the increase and is spreading internationally. The extremists increasingly employ very sophisticated tactics to propagate the climate of fear which is the cornerstone of their strategy. The extremists are also very forensically aware and operate using cell structures similar to those employed by the IRA and other terrorist organisations. In my opinion a wide variety of options are needed to enable the international law enforcement agencies to deal with this increasing brand of terrorism.

Yours sincerely

William Denison

Letter to the Subcommittee on Crime Terrrorism, and Homeland Security from Dr. Amanda Carson Banks, President and CEO, the California Biomedical Research Association

California Biomedical Research Association

1008 Tenth Street, PMB 328, Sacramento, California 95814 Tel 916.558.1515 Fax 916.558.1523
E-mail: info@ca-biomed.org Website: www.ca-biomed.org

May 25, 2006

Members of the House Judiciary Subcommittee on Crime, Terrorism and Homeland Security
2138 Rayburn House Office Building
Washington, DC 20515

RE: SUPPORT: "Animal Enterprise Terrorism Act" – H.R. 4239

On behalf of the Board of Directors and the 86 member facilities of the California Biomedical Research Association, I urge you to support the Animal Enterprise Terrorism Act.

While the intent of the ***original*** Animal Enterprise Protection Act of 1992 was to discourage the unlawful disruption of commerce involving animals, as a legal instrument, it has not been particularly effective – ***no one has been prosecuted*** under the provisions of the 1992 Act since it became law, evidence of its ***ineffectiveness*** as a prosecutorial tool.

There have been, however, COUNTLESS violations of the Act since 1992. In fact, the level of animal enterprise terrorism in the U.S. has dramatically increased since 1992. The FBI has named the Animal Liberation Front and the Earth Liberation Front the top domestic terrorists threat in the U.S. and the Animal Liberation Front (ALF) and the Earth Liberation Front (ELF) alone are responsible for more than 600 criminal acts in the United States since 1996.

The Animal Enterprise Terrorism Act (H.R. 4239) now address key concerns of law enforcement who have waged an uphill battle to prevent, investigate, and prosecute acts of animal rights terrorism. The bills in both the house and the senate amend the Animal Enterprise Protection Act and enhance the effectiveness of the Department of Justice's response to recent trends in the animal rights terrorist movement. H.R. 4239 gives ***better*** tools to law enforcement to help catch eco-terrorists, and provide for stronger sentences to lawbreakers who target ownership and/or use of animals or use of property that interacts with animal habitat, and give strong protections to medical researchers, as well as loggers, miners, fishermen, retailers, labs, farms, animal owners and caretakers, ranchers, processing plants, property owners, and anyone who does business with a company that finds itself with the unwanted attention of eco-terrorists.

Specially, H.R. 4239:

- Broadens the definition of "animal enterprise" to include a commercial enterprise that uses or sell animals or animal products for profit or otherwise including animal shelters, breeders, pet stores, and furriers.

- Addresses the "tertiary targeting" or "third party targeting" system used by animal terrorists by prohibiting the intentional damaging of property of a person or entity having a connection to, relationship with, or transactions with an animal enterprise. Previously, only the law covered the animal enterprise itself.
- Adds penalties for threats, bodily injury, conspiracy, and attempts to conspire • Prohibit veiled threats to individuals and their families. It prohibits intentionally placing a person in reasonable fear of death or serious bodily injury to that person or their family because of their relationship with an animal enterprise.
- Increases penalties for intentionally causing economic disruption or damage and for intentionally causing a person bodily injury or intentionally placing a person in reasonable fear of death or bodily injury.

Currently, California is the nation's leader in the biosciences, with more firms and employment in these industries than any other state. California is also the nation's leader in the private and academic research and development sector that drives these industries. Protecting these important facilities, and the researchers, scientists, and other employees and their facilities is critical. The Animal Enterprise Terrorism Act is just what is needed. In fact, a recent report by the Association of the British Pharmaceutical Industry (ABPI) found that damage in 2005 to company, personal, and public property by animal extremists in the UK was nearly ***halved*** compared with 2004, while abusive or threatening messages dropped by a third. The ABPI's figures show that in 2005 there were 85 instances of damage to property, compared with 177 the previous year. The number of 'home visits' by activists to those involved in research dropped to 57 in 2005, compared to 179 in the previous year. The report credits these drops to the deterrent effect of the new Serious Organized Crime Act, which became law last July. This is strong evidence that the Animal Enterprise Act can and will be successful here if it is passed.

California is also the nation's leader in animal enterprise terrorism. Facilities and researchers are listed on "hit lists", facilities are continuously targeted for "direct actions", and the homes of individuals are visited and vandalized without consequence. I attach an incomplete list of "direct actions" from California over the years. (For security and privacy reasons, this list is not all-inclusive.) Please help our medical researches, scientists, facilities, and their families by supporting this much-needed act.

Thank you for your time and consideration. As always, we appreciate the great work you do representing us in Washington, DC. Please do not hesitate to contact me with any questions or if you would like additional information.

Sincerely,

Amanda Carson Banks, PhD, CFRE
President/CEO

LETTER TO THE U.S. COMMITTEE ON ANIMAL RIGHTS EXTREMISM FROM MRS. WENDY BANTIN

To United States Committee on Animal Rights Extremism

My family have endured a hate campaign by animal rights extremists for 10 years. My husband is involved with the supply of animals for laboratories and as such we are front line targets. Our three children have grown up under police protection and, for their own safety, have had significant restrictions placed upon their freedoms.

When the campaign against us started we believed that the law would protect us but we soon found out that legislation was inadequate and where it could be applied the courts were reluctant to do so. This failure to reach successful prosecutions discouraged the police to charge extremists and the UK has found itself now with a huge problem in dealing with this issue. Too much attention has been given to the Human Rights of the extremists to express their views while our rights to live and work in peace within the requirements of the law have come a poor second.

We continue to suffer attacks not only on ourselves but on those who dare to associate with us or supply the company. The actions do not necessarily have to be violent but the psychological effects we suffer with the constant barrage of implied threats can be intolerable.

This secondary, tertiary and even quaternary targeting has proved to be extremely successful for the animal rights extremists. It effectively aims to cut off the lifeblood to an organization. I'm sure you will have been made aware of the horrific campaign against a guinea pig breeder in Staffordshire who, after many years of standing up to these people in spite of the many attacks on friends, family and business associates, finally could take no more when a family grave was desecrated and the body stolen.

In recent months legislation in the UK has been extended but remains untested in the courts. Enforcement is inconsistent as the law enforcement agencies attempt to interpret and apply it. Meanwhile families like mine continue to live within a climate of fear, not knowing if our livelihoods can survive long enough for the law to become effective.

I hope you can find a way in the United States to counteract this form of terrorism and that we can learn from each other.

Yours sincerely

Wendy Bantin

TOP 20 LIST OF ILLEGAL ACTIONS BY ANIMAL AND ECO-TERRORISTS, 1996–2006, COMPLIED BY THE FOUNDATION FOR BIOMEDICAL RESEARCH

Top 20* List of Illegal Actions by Animal and Eco-Terrorists 1996-2006

IN ORDER OF SEVERITY

1. San Diego Arson
In August 2003, the Earth Liberation Front (ELF) claimed responsibility for setting fire to a new housing development near San Diego – to protest sprawl and development. The planned La Jolla Crossroads complex contemplated 1,500 housing units, including low-income and market-rate rental apartments, as well as condominiums. A banner reading "If you build it, we will burn it, the ELFs are mad," was found at the crime scene, and an e-mail sent to The San Diego Union-Tribune the day of the arson said the banner "is a legitimate claim of responsibility by the Earth Liberation Front." Damage was assessed at more than $50 million.

2. Bombing at Chiron Corporation
In August 2003, two incendiary devices exploded at Chiron Corporation in Northern California. The explosions occurred in the early morning hours causing property damage, but no injuries to people. Chiron Corp. had been a target of escalating harassment and actions due to its alleged connection to Huntington Life Sciences (HLS). There is a $50,000 reward offered for information leading to the arrest of Daniel Andreas San Diego, who authorities believe was involved in setting these bombs as well as the bombing of Shaklee Corp, which was also targeted for its connection to HLS.

3. Bombing at Shaklee Corporation
In September 2003, an incendiary device exploded at Shaklee Corporation in Northern California. Shaklee is a subsidiary of the giant Japanese conglomerate Yamanouchi Holdings Group, which has been a major target of animal activist groups due to its alleged connection to HLS. The bomb was constructed with nails, to create shrapnel, and other readily available materials. No one was injured, and there is a $50,000 reward offered for information leading to the arrest of Daniel Andreas San Diego, who authorities believe to be involved in setting this bomb as well as the bombings at Chiron Corp., which was also targeted for its connection to HLS.

4. Razor Blade Mailing
In October 1999, an animal extremist group calling itself "the justice department" released a communiqué warning that over 80 packages containing razor blades had been mailed to American researchers studying non-human primates. Seven such envelopes were received by research facilities in various parts of the country.

5. Break-in at University of Iowa
In November 2004 the Animal Liberation Front (ALF) claimed responsibility for breaking into and damaging two science buildings at the University of Iowa. The intruders destroyed research, vandalized equipment and stole 400 lab animals. Both buildings were closed for the rest of the semester, and classes relocated out of safety considerations since chemicals were spilled during the break-in and Hazmat teams were required to close the buildings to the public. Damage was assessed at $400,000.

6. Vail Ski Resort Arson
In October 1998, ELF claimed responsibility for an arson fire that caused massive destruction at Vail Ski Resort in October of 1998. The fire destroyed a restaurant, picnic facility, utility building and four ski lifts. An ELF communiqué said the fire was set to protest the proposed expansion of the resort and the declining habitat of lynx. Damage was assessed at $12 million.

7. Break-in, Theft and Destruction at University of Minnesota
In April 1999, during World Week for Animals in Labs, ALF claimed responsibility for breaking into the University of Minnesota and stealing 27 pigeons, 48 mice, 36 rats and five salamanders from Elliot Hall where Alzheimer's and Parkinson's disease research was being conducted. The Lyons Research Building was also broken into and extensive damage was done to equipment, computers and research data and video tapes. Damage was assessed at over $2 million.

8. Break-in, Theft and Destruction at Michigan State University
In December 1999, ELF claimed responsibility for setting an arson fire at the Agriculture Hall at Michigan State University. The blaze destroyed property and years of accumulated research on genetically engineered crops. Catherine Ives, a target in the raid, lost academic records, lecture notes, slide presentations, books, and her passport in the fire. Damage was assessed at more than $1 million.

9. Yacht Damaged
In July of 2001, animal extremists calling themselves "Pirates for Animal Liberation" claimed responsibility for sneaking onto the property of a Bank of New York executive and drilling holes into the side of his boat, sending it out to sea to sink.

10. Break-in, Theft and Destruction at Louisiana State University
In September 2003, ALF claimed responsibility for breaking into LSU's Inhalation Toxicology Research lab, destroying computers and other lab equipment needed to study smoke and other toxins that can contribute to cardiovascular and respiratory disease. Damage was assessed at $250,000. The FBI later offered a $20,000 reward for information leading to the arrest of those responsible.

11. Another attack on Louisiana State University
In September 2004, ALF claimed responsibility for breaking into the Louisiana State University School of Medicine, destroying computers and research equipment. The letters A – L – F were spray painted across a glass partition and red paint was splashed throughout the laboratory. Damage was assessed at $200,000 to $300,000.

12. Two More Yachts Damaged
In August of 2005, animal extremists vandalized two yachts at a Long Island Yacht club. The club was targeted because its membership included employees of Carr Securities, which was trading stock in HLS at the time.

13. Planned Holding Facility Arson
In May 2005 animal extremists set an arson fire at Peonyland, a greenhouse owned by a Bucks County, Pennsylvania businessman who had recently applied for a building permit for the construction of a holding facility for non human primates needed in biomedical research. Peonyland's greenhouse was burned to the ground, destroying plants, and vandalizing cars and other personal property on the same evening the owner was attending a public hearing on his application, which he later withdrew.

14. Break-in, Theft and Destruction at Brigham Young University
In July 2004, the letters "ALF" were found spray-painted in seven locations at Brigham Young University's agriculture center, near a recycling building to which firefighters had been called to extinguish a suspicious fire. This was the third incident that BYU attributed to ALF in a six week period during which a barn was broken into, animals and equipment stolen. Damaged was assessed at $30,000.

15. Boise Cascade Arson
In December 1999, ELF claimed responsibility for an arson fire that destroyed Boise Cascade's regional headquarters in Monmouth, Oregon. A communiqué from the ELF stated: Early Christmas morning elves left coal in Boise Cascade's stocking. Four buckets of diesel gas with kitchen timer delay destroyed their regional headquarters." Damage was assessed at $1 million.

16. Break-in, Theft and Destruction at Western Washington University
Twice within a two month period in 1999, labs at Western Washington University were broken into. Offices were damaged, equipment and research data destroyed and lab animals stolen. Damage to the lab assessed at tens of thousands of dollars.

17. University of Washington Arson
In May 2001, a fire was set at the Center for Urban Horticulture at the University of Washington in Seattle. The arson was attributed to the ELF, which objected to the plant research being conducted. Damage was assessed at $5 million.

18. Bombing at B&K Universal

In February 2000, ALF claimed responsibility for placing four incendiary devices at the warehouse of B&K Universal, a supplier of products to the animal research community. Several trucks were damaged.

19. Break-in, Theft and Destruction at UCSF

In April 1999, animal extremists broke into three research labs at the University of California, San Francisco, shattering glass, damaging equipment and destroying research. Four mice were stolen. A group of World Laboratory Animal Week demonstrators broke into the facility and confronted a researcher. Three people were arrested. Damages were assessed at thousands of dollars.

20. Arson at USDA Research Lab

In June 1998, ALF and ELF claimed responsibility for setting fires at two USDA animal research buildings in Washington State. A communiqué referred to "a bonfire at facilities which make it a daily routine to kill and destroy wildlife. Structural damaged was assessed at more than $400,000.

***21. HLS Employee Beaten**

In the UK, in February 2001, the managing director of HLS was brutally attacked by three baseball-bat wielding thugs in the garage of his own home. Brian Cass was hospitalized by the attack.

Compiled from various sources by the Foundation for Biomedical Research, 2006.

IN CHRONOLOGICAL ORDER

1. Two More Yachts Damaged
In August of 2005, animal extremists vandalized two yachts at a Long Island Yacht club. The club was targeted because its membership included employees of Carr Securities, which was trading stock in HLS at the time.

2. Planned Holding Facility Arson
In May 2005 animal extremists set an arson fire at Peonyland, a greenhouse owned by a Bucks County, Pennsylvania businessman who had recently applied for a building permit for the construction of a holding facility for non human primates needed in biomedical research. Peonyland's greenhouse was burned to the ground, destroying plants, and vandalizing cars and other personal property on the same evening the owner was attending a public hearing on his application, which he later withdrew.

3. Break-in at University of Iowa
In November 2004 the Animal Liberation Front (ALF) claimed responsibility for breaking into and damaging two science buildings at the University of Iowa. The intruders destroyed research, vandalized equipment and stole 400 lab animals. Both buildings were closed for the rest of the semester, and classes relocated out of safety considerations since chemicals were spilled during the break-in and Hazmat teams were required to close the buildings to the public. Damage was assessed at $400,000.

4. Another attack on Louisiana State University
In September 2004, ALF claimed responsibility for breaking into the Louisiana State University School of Medicine, destroying computers and research equipment. The letters A – L – F were spray painted across a glass partition and red paint was splashed throughout the laboratory. Damage was assessed at $200,000 to $300,000.

5. Break-in, Theft and Destruction at Brigham Young University
In July 2004, the letters "ALF" were found spray-painted in seven locations at Brigham Young University's agriculture center, near a recycling building to which firefighters had been called to extinguish a suspicious fire. This was the third incident that BYU attributed to ALF in a six week period during which a barn was broken into, animals and equipment stolen. Damaged was assessed at $30,000.

6. Break-in, Theft and Destruction at Louisiana State University
In September 2003, ALF claimed responsibility for breaking into LSU's Inhalation Toxicology Research lab, destroying computers and other lab equipment needed to study smoke and other toxins that can contribute to cardiovascular and respiratory disease. Damage was assessed at $250,000. The FBI later offered a $20,000 reward for information leading to the arrest of those responsible.

7. Bombing at Shaklee Corporation
In September 2003, an incendiary device exploded at Shaklee Corporation in Northern California. Shaklee is a subsidiary of the giant Japanese conglomerate Yamanouchi

Holdings Group, which has been a major target of animal activist groups due to its alleged connection to HLS. The bomb was constructed with nails, to create shrapnel, and other readily available materials. No one was injured, and there is a $50,000 reward offered for information leading to the arrest of Daniel Andreas San Diego, who authorities believe to be involved in setting this bomb as well as the bombings at Chiron Corp., which was also targeted for its connection to HLS.

8. Bombing at Chiron Corp.
In August 2003, two incendiary devices exploded at Chiron Corporation in Northern California. The explosions occurred in the early morning hours causing property damage, but no injuries to people. Chiron Corp. had been a target of escalating harassment and actions due to its alleged connection to Huntington Life Sciences (HLS). There is a $50,000 reward offered for information leading to the arrest of Daniel Andreas San Diego, who authorities believe was involved in setting these bombs as well as the bombing of Shaklee Corp., which was also targeted for its connection to HLS.

9. San Diego arson
In August 2003, the Earth Liberation Front (ELF) claimed responsibility for setting fire to a new housing development near San Diego – to protest sprawl and development. The planned La Jolla Crossroads complex contemplated 1,500 housing units, including low-income and market-rate rental apartments, as well as condominiums. A banner reading "If you build it, we will burn it, the ELFs are mad," was found at the crime scene, and an e-mail sent to The San Diego Union-Tribune the day of the arson said the banner "is a legitimate claim of responsibility by the Earth Liberation Front." Damage was assessed at more than $50 million.

10. Yacht Damaged
In July of 2001, animal extremists calling themselves "Pirates for Animal Liberation" claimed responsibility for sneaking onto the property of a Bank of New York executive and drilling holes into the side of his boat, sending it out to sea to sink.

11. University of Washington Arson
In May 2001, a fire was set at the Center for Urban Horticulture at the University of Washington in Seattle. The arson was attributed to the ELF, which objected to the plant research being conducted. Damage was assessed at $5 million.

12. Bombing at B&K Universal
In February 2000, ALF claimed responsibility for placing four incendiary devices at the warehouse of B&K Universal, a supplier of products to the animal research community. Several trucks were damaged.

13. Break-in, Theft and Destruction at Michigan State University
In December 1999, ELF claimed responsibility for setting an arson fire at the Agriculture Hall at Michigan State University. The blaze destroyed property and years of accumulated research on genetically engineered crops. Catherine Ives, a target in the raid,

lost academic records, lecture notes, slide presentations, books, and her passport in the fire. Damage was assessed at more than $1 million.

14. Boise Cascade Arson

In December 1999, ELF claimed responsibility for an arson fire that destroyed Boise Cascade's regional headquarters in Monmouth, Oregon. A communiqué from the ELF stated: Early Christmas morning elves left coal in Boise Cascade's stocking. Four buckets of diesel gas with kitchen timer delay destroyed their regional headquarters." Damage was assessed at $1 million.

15. Western Washington University vandalized twice

Twice within a two month period in 1999, labs at Western Washington University are struck. Offices and experiments are destroyed and lab animals were stolen. Damage was assessed at tens of thousands of dollars.

16. Razor blade mailing

In October 1999, an animal extremist group calling itself "the justice department" released a communiqué warning that over 80 packages containing razor blades had been mailed to American researchers studying non-human primates. Seven such envelopes were received by research facilities in various parts of the country.

17. Break-in, Theft and Destruction at University of Minnesota

In April 1999, during World Week for Animals in Labs, ALF claimed responsibility for breaking into the University of Minnesota and stealing 27 pigeons, 48 mice, 36 rats and five salamanders from Elliot Hall where Alzheimer's and Parkinson's disease research was being conducted. The Lyons Research Building was also broken into and extensive damage was done to equipment, computers and research data and video tapes. Damage was assessed at over $2 million.

18. Break-in, Theft and Destruction at UCSF

In April 1999, animal extremists broke into three research labs at the University of California, San Francisco, shattering glass, damaging equipment and destroying research. Four mice were stolen. A group of World Laboratory Animal Week demonstrators broke into the facility and confronted a researcher. Three people were arrested. Damages were assessed at thousands of dollars.

19. Vail Ski Resort Arson

In October 1998, ELF claimed responsibility for an arson fire that caused massive destruction at Vail Ski Resort in October of 1998. The fire destroyed a restaurant, picnic facility, utility building and four ski lifts. An ELF communiqué said the fire was set to protest the proposed expansion of the resort and the declining habitat of lynx. Damage was assessed at $12 million.

20. Arson at USDA Research Lab

In June 1998, ALF and ELF claimed responsibility for setting fires at two USDA animal research buildings in Washington State. A communiqué referred to "a bonfire at facilities which make it a daily routine to kill and destroy wildlife. Structural damaged was assessed at more than $400,000.

***21.** In the UK in February 2001, Managing Director of HLS Brian Cass was brutally attacked and hospitalized by three baseball-bat wielding thug in the garage of his own home. The attack was meant to send a message against HLS and animal research.

Compiled from various sources by the Foundation for Biomedical Research, 2006.

NEWS ARTICLE, *ANIMAL RESEARCH DOES NOT VALIDATE TRESPASS*

THURSDAY, MAY 1, 2003 • THE DAILY BRUIN 9

Animal research does not validate trespass

By John Schlag and Madeleine Schlag-Rey

In the Daily Bruin News article, "Animal rights activists protest against UCLA researchers" (April 25), Nicolas Taborek reports protesters demonstrated outside the homes of UCLA animal researchers and one of the activists, Erica Sutherland, "believes the demonstrations were effective."

We are two of these "animal researchers." And, indeed, on April 21, around 10:15 p.m., our home was attacked allegedly by a large group of animal activists, wearing masks to hide their identities. They were shouting, throwing stones, kicking and banging so hard on the entrance door of our home that we feared they wanted to demolish it. Not only were they trespassing on our private road, but they trespassed on our property. They broke the glass window panel of the side door to our garage. A rock, thrown from the street, made a hole in the screen door – fortunately not the glass door situated a few feet from our dining table. In the house next door, a four-year-old little girl was filled with terror.

What is the result of these acts of intimidation? Immediately, we had calls and e-mails from our neighbors, some of whom had seen the activists and/or the fliers accusing us. Their reactions were very refreshing. Some of our neighbors even became interested in the research we were doing at UCLA. All were very supportive. Here is a sample of the messages received:

"Please take comfort in knowing that we are outraged by the PETA's (People for the Ethical Treatment of Animals) actions on the road last night. We support the work that you do for the good of mankind, and we stand ready to help you if ever we can be of service."

"As your neighbors, we are always there for you."

"Frightening children in their sleep is simply not to happen again."

We wish to address our own message to the PETA – regardless of the cause, criminal trespass and the infliction of fear, emotional distress and terror cannot be countenanced in our society. I understand that you believe you are fighting for the benefit of the defenseless. However, the behavior manifested by black-cloaked marauders at 10 p.m. is criminal terrorism, trespass and harassment. We will pursue civil and criminal remedies if a repetition is considered.

The countless letters from our neighbors do not support Erica Sutherland's contention that "the demonstrations were effective." On the contrary, these letters prove that terrorism does not pay.

Animal activists may be outraged because so few people take their arguments seriously, but it is not a reason to rely on aggression as an act of desperation.

The next morning, one of us (John) gave two lectures to a class of first-year medical students and, of course, the material included findings obtained from animal research. Whether we like it or not, the substantial and accelerating progress medicine has made since the second half of the 19th century is due, in major part, to experimental research on animals. All physicians are trained on the basis of this research.

Think about this the next time you see your doctor because at that time you are taking advantage of what he or she has learned from animal research. Remember that the medications prescribed for you have been tested on animals for the sake of your own safety. What's more, not only human medicine, but veterinary medicine, benefits from animal research.

John Schlag is a neurobiology professor at UCLA. Madeleine Schlag-Rey is a neurobiology researcher at UCLA.